Y0-EHC-774

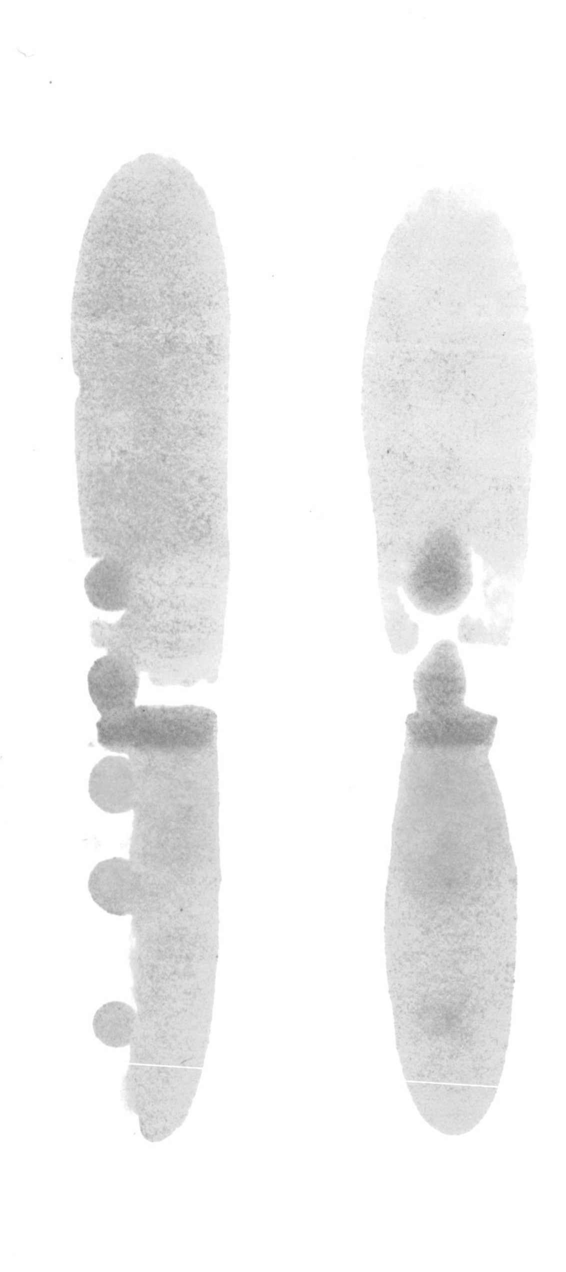

VOLUME 495 JANUARY 1988

THE ANNALS

of The American Academy *of* Political
and Social Science

TELESCIENCE: SCIENTIFIC COMMUNICATION IN THE INFORMATION AGE

Special Editor of this Volume

MURRAY ABORN

Senior Scientist
Division of Social and Economic Science
National Science Foundation
Washington, D.C.

SAGE PUBLICATIONS *NEWBURY PARK BEVERLY HILLS LONDON NEW DELHI*

THE ANNALS

ERICA GINSBURG, *Assistant Editor*

Editorial Office: 3937 Chestnut Street, Philadelphia, Pennsylvania 19104.

For information about membership (individuals only) and subscriptions (institutions), address:*

SAGE PUBLICATIONS, INC.

2111 West Hillcrest Drive
Newbury Park, CA 91320

275 South Beverly Drive
Beverly Hills, CA 90212

From India and South Asia, write to:

SAGE PUBLICATIONS INDIA Pvt. Ltd.
P.O. Box 4215
New Delhi 110 048
INDIA

From the UK, Europe, the Middle East and Africa, write to:

SAGE PUBLICATIONS LTD
28 Banner Street
London EC1Y 8QE
ENGLAND

SAGE Production Editors: JANET BROWN and ASTRID VIRDING

**Please note that members of The Academy receive THE ANNALS with their membership.*

Library of Congress Catalog Card Number 86-063790
International Standard Serial Number ISSN 0002-7162
International Standard Book Number ISBN 0-8039-2938-2 (Vol. 495, 1988 paper)
International Standard Book Number ISBN 0-8039-2937-4 (Vol. 495, 1988 cloth)
Manufactured in the United States of America. First printing, January 1988.

The articles appearing in THE ANNALS are indexed in *Book Review Index; Public Affairs Information Service Bulletin; Social Sciences Index; Monthly Periodical Index; Current Contents; Behavioral, Social Management Sciences;* and *Combined Retrospective Index Sets.* They are also abstracted and indexed in *ABC Pol Sci, Historical Abstracts, Human Resources Abstracts, Social Sciences Citation Index, United States Political Science Documents, Social Work Research & Abstracts, Peace Research Reviews, Sage Urban Studies Abstracts, International Political Science Abstracts, America: History and Life,* and/or *Family Resources Database.*

Information about membership rates, institutional subscriptions, and back issue prices may be found on the facing page.

Advertising. Current rates and specifications may be obtained by writing to THE ANNALS Advertising and Promotion Manager at the Newbury Park office (address above).

Claims. Claims for undelivered copies must be made no later than three months following month of publication. The publisher will supply missing copies when losses have been sustained in transit and when the reserve stock will permit.

Change of Address. Six weeks' advance notice must be given when notifying of change of address to insure proper identification. Please specify name of journal. Send change of address to: THE ANNALS, c/o Sage Publications, Inc., 2111 West Hillcrest Drive, Newbury Park, CA 91320.

Origin and Purpose. The Academy was organized December 14, 1889, to promote the progress of political and social science, especially through publications and meetings. The Academy does not take sides in controverted questions, but seeks to gather and present reliable information to assist the public in forming an intelligent and accurate judgment.

Meetings. The Academy holds an annual meeting in the spring extending over two days.

Publications. THE ANNALS is the bimonthly publication of The Academy. Each issue contains articles on some prominent social or political problem, written at the invitation of the editors. Also, monographs are published from time to time, numbers of which are distributed to pertinent professional organizations. These volumes constitute important reference works on the topics with which they deal, and they are extensively cited by authorities throughout the United States and abroad. The papers presented at the meetings of The Academy are included in THE ANNALS.

Membership. Each member of The Academy receives THE ANNALS and may attend the meetings of The Academy. Membership is open only to individuals. Annual dues: $28.00 for the regular paperbound edition (clothbound, $42.00). Add $9.00 per year for membership outside the U.S.A. Members may also purchase single issues of THE ANNALS for $6.95 each (clothbound, $10.00).

Subscriptions. THE ANNALS (ISSN 0002-7162) is published six times annually—in January, March, May, July, September, and November. Institutions may subscribe to THE ANNALS at the annual rate: $60.00 (clothbound, $78.00). Add $9.00 per year for subscriptions outside the U.S.A. Institutional rates for single issues: $10.00 each (clothbound, $15.00).

Second class postage paid at Philadelphia, Pennsylvania, and at additional mailing offices.

Single issues of THE ANNALS may be obtained by individuals who are not members of The Academy for $7.95 each (clothbound, $15.00). Single issues of THE ANNALS have proven to be excellent supplementary texts for classroom use. Direct inquiries regarding adoptions to THE ANNALS c/o Sage Publications (address below).

All correspondence concerning membership in The Academy, dues renewals, inquiries about membership status, and/or purchase of single issues of THE ANNALS should be sent to THE ANNALS c/o Sage Publications, Inc., 2111 West Hillcrest Drive, Newbury Park, CA 91320. *Please note that orders under $25 must be prepaid.* Sage affiliates in London and India will assist institutional subscribers abroad with regard to orders, claims, and inquiries for both subscriptions and single issues.

THE ANNALS

of The American Academy *of* Political *and* Social Science

FORTHCOMING

STATE CONSTITUTIONS IN A FEDERAL SYSTEM
Special Editor: John Kincaid
Volume 496 — March 1988

ANTI-AMERICANISM: ORIGINS AND CONTEXT
Special Editor: Thomas P. Thornton
Volume 497 — May 1988

THE PRIVATE SECURITY INDUSTRY: ISSUES AND TRENDS
Special Editor: Ira A. Lipman
Volume 498 — July 1988

See page 3 for information on Academy membership and purchase of single volumes of **The Annals.**

CONTENTS

BOOK DEPARTMENT CONTENTS

INTERNATIONAL RELATIONS AND POLITICS

AFRICA, ASIA, AND LATIN AMERICA

SOCIOLOGY

ECONOMICS

PREFACE

Science is on the verge of being transformed by one of its own prodigies—information technology. Underlying this technology is the digital computer, whose memory and processing capacities are at the threshold of unprecedented expansion. Already available are new computer architectures composed of parallel processing systems so powerful that they are adding to theorizing and experimentation a third mode of doing science. Over the doorstep are software systems so advanced that they will provide the scientist with automatically generated computer programs tailored to highly specific research tasks, immensely increasing research productivity. Under development are tools for interfacing the researcher with his or her equipment in ways that are much more natural, or user-friendly; intelligent, providing expert assistance; and flexible, being able, among other things, to give instructions by spoken word. Also on the horizon are improvements in glass fiber and laser transmission media, and improved network switching devices, that will enormously increase the ability of computing systems to communicate with each other.

The machine underlying this technology-based revolution in science is also the prime mover of the so-called information age, an epochal event not likely to have escaped the attention of *Annals* readers. While impacts of the information age on society at large have been fairly well publicized, assessments of its impact on the scientific enterprise itself have been relatively restricted. The present volume takes its impetus from that state of affairs.

As the world continues to move ever more deeply into the information age, the role played by science and technology in national and international decision making grows more prominent. In turn, the economic and political implications of information technology are becoming preponderant considerations in technical decision making, as are sociological and psychological factors. Few works in what I shall dub the information-age literature fail to emphasize the cardinal importance of sociological and psychological factors when discussing the infrastructures required for the effective operation of informational systems. The significance of these interconnections for the scientific enterprise stands out in many of the articles composing the present volume.

To return to the major theme of the volume, its purpose is to examine the impingement of the information age on science per se—and, more particularly, on scientific interchange and communication—rather than examine the ultimate social transmutations that may take place. The term "telescience" is adopted for the purpose of conveying the increasingly global scope of scientific intercourse that the information age tends to spur and because it depicts an essential feature of the communicational environment in which science will henceforth be conducted.

"Telescience" is attributed to the National Aeronautics and Space Agency, where it is used to refer to scientific research carried out remotely via computer-based networks, involving collaboration between scientists far removed from each other and often from their physical facilities. This *Annals* volume will treat the term more loosely to include the distancing of researchers from the very process of science and to cover the negative as well as the positive consequences of the shift away from modes and norms that have characterized scientific communication in the past.

This volume aims at supplying the reader with a well-balanced commentary and not so much a state-of-the-art exposition or analytical review, although elements of both styles are to be found among the various articles. The articles themselves have been selected with an eye toward representation insofar as the different fields of science are concerned, while the authors bring to bear expertise stemming from a wide variety of disciplinary backgrounds and knowledge backed by eminent achievements in scientific research and administration.

In presenting the ways in which information technology is affecting science, then, the authors use perspectives from the social sciences, the health sciences, and the physical sciences and draw upon proficiencies acquired through scholarship in chemistry, computer science, economics, history of science, linguistics, mathematics, meteorology, medicine, and sociology. The reader will discover a diversity of viewpoints and a wide range of experiences in the pages that follow, but also a certain coherence emanating from the scientific training of the authors and the connectedness of the topics they were asked to address.

ARTICLES IN THIS VOLUME

James R. Beniger, now at the Annenberg School of Communications, University of Southern California, was asked to lay a historical foundation for the coming of the information age, the intent being to put the impact of the computer on science into historical perspective and relate it not just to other technological developments but to changes in social organization and broad-scale societal functioning as well. In his article, titled "Information Society and Global Science," Dr. Beniger does this—and more!

A similar request was made of Melvin Kranzberg, historian of technology par excellence. In this case, however, the charge was to provide background on the traditional relationship between technology and science and to explain how and why the change we are witnessing in that relationship is likely to affect scientific communication as a social as well as an informational process. Professor Kranzberg analyzes the interplay of forces that have, over time, led to the virtual disappearance of the distinction between science and technology, but he argues that the new relationship is, if anything, more embedded in sociopolitical processes than before. To hammer home a broader point, he opens and closes his analysis with a description of scientists communicating in ways we all recognize as instinctively human—the encroachments of information technology notwithstanding.

The Beniger and Kranzberg articles are foundational in purpose, which is to say that they are meant to provide a sociological context for the volume as a whole and set the stage for the contributions that follow. The next four articles bring into relief some of the costs and benefits associated with information technology's takeover of traditional scientific linkages. Dilemmas exist because science is at one and the same time a personal process and a communal proceeding, while its inherent values can be made to apply to both national and international aspirations.

Richard Jay Solomon, research associate at the Massachusetts Institute of Technology and editor of *International Networks*, elucidates the complex workings of today's and tomorrow's global telecommunications environment and shows us the cultural, sociological, and psychological dark sides of a world dominated by information-transfer technologies. For political entities there looms the loss of sovereignty; for science, the loss of intellectual control.

In an article dealing with the sharing of scientific data, Theodor D. Sterling of the Simon Fraser University School of Computing Science brings to light the dark in the data as well as the dark side of the data-sharing process. He posits two opposing theses: (1) that technological change forces change in scientific communication practices; and (2) that scientists select from among available technologies those that favor established communication practices. He emphasizes and provides evidence in support of the latter.

The problems of containing and controlling the flow of scientific information across national boundaries are delineated and discussed by Dr. Fred W. Weingarten, manager of the Communication and Information Technologies Program of the Office of Technology Assessment, United States Congress, and his colleague, D. Linda Garcia. Their article defines and describes the evolution of information policy and analyzes how concerns in the areas of national security and trade, plus communications regulations in other areas, affect the ability of researchers to exchange scientific knowledge. They warn that unless the community of research scientists becomes more active in contributing to the formulation of well-balanced information policies, the free exercise of scientific inquiry may become impossible.

Completing what may be called the set of issue-oriented articles is the description of a mathematically based method for reducing an age-old barrier to scientific communication—namely, the diversity of languages in which scientific information is conveyed. The problems of linguistic divergence in science are exacerbated in the information age owing to the increasing internationalization of telecommunication networks. Professor Zellig Harris of the University of Pennsylvania and Dr. Paul Mattick, Jr., of New York University demonstrate how information science may one day provide the basis for a unified language of science that can surmount national language differences.

The next five articles are intended to furnish examples of telescience in action, so to speak. These articles are in the nature of case studies, each providing an instance of successful functioning by a major national or international organization.

The set begins with an article by Jacques Tocatlian, director of the United Nations Educational, Scientific, and Cultural Organization's General Information Program. He describes the role played by this arm of the United Nations in laying both a scientific and a policymaking foundation for the free flow of scientific and technical information across national borders.

The following two articles describe organizations employing information technology on an international scale to achieve specific, substantively oriented scientific objectives. The first, written by Professors Lee Rainwater of Harvard University and Timothy M. Smeeding of Vanderbilt University, explains the purpose and operating principles of the Luxembourg Income Study, a cooperative international research project functioning by means of a computerized telecommunications network that provides access to a centrally located multinational data base of income and welfare statistics. The second article represents a shift in substance from social science to geoscience and is written by Dr. Eugene W. Bierly, director of the Division of Atmospheric Sciences, National Science Foundation. Dr. Bierly describes how many different types of national and international organizations it takes to collect, integrate, and disseminate via telecommunications technology the information that must be communicatively linked in order to advance knowledge of global and regional climatological phenomena. He argues that sociopolitical and administrative complexities notwithstanding, the mainspring of the system is science, and the system will function effectively only if there is continued progress in science.

The last two articles in the case-study set deal with scientific organizations whose activities are more circumscribed geographically than the ones discussed in the three preceding articles but whose operations represent telescience in action nonetheless. The first is written by Dr. Harold M. Schoolman, deputy director for research and education, National Library of Medicine, in collaboration with Dr. Donald A.B. Lindberg, director of the National Library of Medicine. This article reviews the role of the Library in the development of medical informatics, a new multidisciplinary field that attempts to provide a scientific basis for the use of automated information systems in biomedicine. The second article, by Murray Aborn and Alvin I. Thaler of the National Science Foundation, provides a historical backdrop for and discusses the logic behind a programmatic initiative called Experimental Research in Electronic Submission. This initiative is being undertaken to prepare the way for utilizing emerging information technologies to facilitate and enhance the development, proposing, review, and funding of scientific research.

In an epilogue, the special editor of this volume mulls over the prospect of a diminishing role for the scientist as the information age envelops the process of science as such. He is reminded of a review elsewhere of recent popular books that attempt to employ the paradoxes of quantum physics to bolster particular spiritual beliefs. The writers of the review observe that an earlier age, the age of Copernicus, banished man from the center of the universe and that "science has steadily whittled us down to size ever since."[1] In the final article, titled "Machine Cognition and the Downloading of Scientific Intellect," the editor proposes that science itself is about to be whittled down to size.

MURRAY ABORN

1. Robert P. Crease and Charles C. Mann, "Physics for Mystics," *Sciences,* 27(4):50-57 (July-Aug. 1987).

ANNALS, *AAPSS*, **495**, January 1988

Information Society and Global Science

By JAMES R. BENIGER

ABSTRACT: In many advanced industrial countries, much of the labor force works at informational tasks, while wealth comes increasingly from informational goods and services. But why? Answers can be found in the Control Revolution, an abrupt change in the technological and economic arrangements by which information is collected, stored, processed, and communicated that began in the middle of the nineteenth century. The early history of computers in science can be understood from this perspective, as can telematics, the recent convergence—through digitalization—of information-processing and communication technologies. Telematics does not necessarily follow from advanced industrialization because new control technologies present societal choices subject to political control. Problems of control in an interorganizational system like science are considered, with special attention to the role of generalized symbolic media in distributing system status and exchange authority across organizational boundaries. Potential control problems are foreseen in a telematic basis for the system of science, in which computerization has tended to minimize exchange authority, thereby threatening both interorganizational incentives and macro-level control.

James R. Beniger graduated magna cum laude in history from Harvard. He has an M.S. in statistics and an M.A. and Ph.D. in sociology from the University of California at Berkeley. His most recent book, The Control Revolution: Technological and Economic Origins of the Information Society, *won the Association of American Publishers award for the most outstanding book in the social and behavioral sciences in 1986. Currently he is associate professor at the Annenberg School of Communications, University of Southern California.*

To say that the advanced industrial world has become an information society has already become a cliché. In not only the United States but also Canada, Western Europe, and Japan, the bulk of the labor force now works primarily at informational tasks, including such readily identifiable ones as systems analysis and computer programming, while wealth comes increasingly from informational goods such as microprocessors and from informational services such as data processing.

Both the timing and direction of this great societal transformation can be measured using U.S. labor force statistics. In 1880, fewer than 7 percent of American workers produced informational goods and services, compared to the nearly 45 percent in agriculture. By far the fastest growth in information work came during the 1880s, when the sector nearly doubled to more than 12 percent of the work force. By 1930, the information sector had doubled again to occupy a quarter of all labor, compared to 35 percent in other industry and 20 percent each in other services and in agriculture.

Today, with many of those born in 1930 still in the labor force, America's information sector has doubled once again. Roughly half of us now earn our living from informational products, compared to less than 30 percent from other services, 20 percent from other industry, and scarcely 2 percent from agriculture.[1] The manufacture of noninformational goods—the so-called smokestack industries—once the backbone of the American economy, may employ fewer than 15 percent of our workers by the end of this decade, even as farm work all but disappears.

AN EMERGING GLOBAL SECTOR

Parallel developments have transformed the economies of at least a dozen advanced industrial countries. A Japanese study finds that informational goods and services accounted for 35.4 percent of that nation's gross national product (GNP) in 1979, up from 21.3 percent in 1960; the information sector of Japan's work force increased to 37.7 from 21.3 percent over the same period.[2] Interpolating from comparable U.S. data,[3] information rose to 34.3 from 29.6 percent of GNP and to 53.1 from 42.1 percent of the American work force during the same years. At least by the metric of GNP—though not labor force composition—the Japanese data suggest that that nation passed the United States as the leading information society sometime in 1974 or 1975.

A study by the Organization for Economic Cooperation and Development of its member nations in 1978 and 1979 finds the information sector ranging from 14.8 percent of GNP in Australia to 24.8 percent in France and the United States.[4] These percentages include only

1. James R. Beniger, *The Control Revolution: Technological and Economic Origins of the Information Society* (Cambridge, MA: Harvard University Press, 1986), pp. 21-24.

2. Seisuke Komatsuzaki et al., *An Analysis of Information Economy in Japan from 1960 to 1980* (Tokyo: Research Institute of Telecommunications and Economics, n.d.).

3. Michael Rogers Rubin and Mary Taylor Huber, *The Knowledge Industry in the United States, 1960-1980* (Princeton, NJ: Princeton University Press, 1986), pp. 19, 196.

4. Organization for Economic Cooperation and Development, *Information Activities, Electronics and Telecommunications Activities, Impact on Employment, Growth, and Trade* (Paris: Organization for Economic Cooperation and Development, 1981).

the primary information sector, that is, only the informational goods and services sold directly to the market, excluding the roughly equal dollar amount produced and consumed internally by noninformation firms and by governments. This study also measures the spread of information work through the labor forces of advanced industrial countries, using data for individual years between 1970 and 1978: 27.5 percent in Finland, 29.6 in Japan, 32.1 in France, 32.2 in Austria, 33.2 in West Germany, 34.9 in Sweden, 35.6 in Great Britain, 39.9 in Canada, and 41.1 in the United States.

These and similar studies from three continents have identified a historic transformation of human society: the emergence of information as a major independent sector of the global economy. Unlike all other societies we know, in some 50,000 years of human history, a dozen or so nations now depend on informational goods and services more than on hunting and gathering, agriculture and mining, or noninformational manufacturing and commerce. The processing of matter and energy, it would seem, has begun to be overshadowed by the processing of information.

But why? Among the multitude of things that human beings value, why should it be information, embracing both goods and services, that has come to dominate the world's largest and most advanced economies? And why now? Information plays an important role in all human societies, after all—why in only this century should it emerge as a distinct and critical commodity?

THE CONTROL REVOLUTION

Answers lie in what I call the Control Revolution,[5] a concentration of abrupt changes in the technological and economic arrangements by which information is collected, stored, processed, and communicated and through which formal or programmed decisions might effect societal control. From its origins in the later decades of the nineteenth century, the Control Revolution has continued to this day, sustained—in its more recent stages—by the appearance of business computers in the 1950s, microprocessors in the 1970s, and personal computers in the 1980s, as well as by countless other technological developments.

To glimpse the future course of this change, the technological counterpart to the transformation of the American labor force, it might be useful to reflect on its initial cause. The Control Revolution began as a response to rapid industrialization after 1830 and to the resulting crisis in control of the material economy.[6] Before the application of steam power, even the largest and most developed economies ran literally at a human pace, with processing speeds enhanced somewhat by draft animals and by wind and water power, but still well within the information-processing capabilities of individual human brains to control. System-level control could be maintained by relatively flat bureaucratic structures.

By far the greatest impact of industrialization, from the perspective of societal control, was to speed up the entire material economy, the system for the extraction, processing, and distribution of commodities from environmental input to final consumption. Almost overnight, with the harnessing of steam power, material flows could be moved ten to a hundred times faster, day and night and in virtually any weather. This

5. Beniger, *Control Revolution,* esp. pt. 3.

6. Ibid., chap. 6.

brought widespread breakdowns of control: fatal train wrecks, misplacement of freight cars, loss of shipments, inability to maintain high rates of inventory turnover. What began as a crisis of safety on the railroads in the early 1840s spread to distribution—commission trading and wholesaling—by the 1850s, to production—rail mills and other metal-making and metalworking industries—in the late 1860s, and finally to the marketing of vast outputs of continuous-processing industries—flour, soap, cigarettes, matches, canned goods, and photographic film—in the early 1880s.

Even the word "revolution" seems barely adequate to describe what followed: the development, within the span of a single lifetime, of virtually every basic information-processing and communication technology still in use a century later. These included telegraphy and rotary power printing (1840s), postage stamps and a transatlantic cable (1850s), paper money and modern bureaucracy (1860s), the typewriter, telephone, and switching exchange (1870s), punch-clock, cash register, and Linotype (1880s), motion pictures, magnetic tape recording, and four-function calculators (1890s), and electronic broadcasting (1900s). Just as the Industrial Revolution had marked a historical discontinuity in the ability to harness energy, the Control Revolution marked a similarly dramatic leap in the ability to exploit information.

But why does the Control Revolution continue to this day, a century and a half since the onset of rapid industrialization? Several forces seem to sustain its momentum. Energy utilization, processing speeds, and control technologies have continued to coevolve in a positive spiral, advances in any one factor causing—or at least enabling—improvements in the other two. Additional energy has increased not only the speed of material processing and transportation but also their volume and predictability, which in turn have further increased both the demand for control and returns on new applications of information technology. Information processing and flows themselves need to be controlled, so that information technologies must continue to be applied at higher and higher layers of control—certainly an ironic twist to a Control Revolution.

ENTER THE COMPUTER

Only through appreciation of the Control Revolution, I believe, can we hope to understand otherwise mysterious aspects of the history of the computer, such as why so many of the machine's major components had been anticipated by mid-nineteenth century. As early as 1833, Charles Babbage had designed his steam-powered Analytical Engine with the essential components of a digital computer: punch-card input and programming, internal memory ("store"), a central processing unit ("mill"), and output to be printed or set into type. Far from the ideas of a visionary mathematician, as they are often portrayed, Babbage's design followed by only six years his work on control of the British postal system and by one year the publication of his *On the Economy of Machinery and Manufactures,* a pioneering treatise on industrial control based on exhaustive empirical study—later reprinted as the first text on operations research. Six years after beginning work on his Analytical Engine, Babbage turned back to the crisis of industrial control in a series of studies of the great Western Railway.[7]

A century later in 1937, Howard Aiken, a former Westinghouse engineer

7. Anthony Hyman, *Charles Babbage: Pioneer of the Computer* (Princeton, NJ: Princeton University Press, 1982).

teaching applied mathematics at Harvard, drafted a proposal arguing that scientists needed more powerful computing.[8] Inspired by Babbage's work on industrial control, Aiken included as an example the purposive monitoring and control of the material economy, what he termed the "science of mathematical economy"; his Harvard colleague, economist Wassily Leontief, had published a formal theory of input-output analysis toward the same goal earlier in the year, work that would culminate in a Nobel Prize in 1973. As an appendix to the proposal that, funded by IBM, would in six years yield the electromechanical Mark I, Aiken included a gloss of Babbage's 1833 design—proof enough of the intellectual continuity of the Control Revolution over the intervening century. Indeed, Mark I was in many ways inferior to Babbage's design: the new machine lacked a differentiated processing structure and any general-purpose central processing unit.

THE ORIGINS OF SCIENTIFIC COMPUTING

The electronic numerical integrator and calculator (ENIAC), a fully functional electronic machine that ran its first program in November 1945 at the Moore School of Electrical Engineering, University of Pennsylvania, is often cited as the first scientific computer.[9] If we view computing as only one of several major developments in the emergence of the information society, however, ENIAC appeared closer to the midpoint than to the beginning. The machine ought to be seen as the culmination of work on generalized information-processing technology begun with the Control Revolution and interrupted by World War II.[10] Consider the intellectual as well as technological momentum the Control Revolution had gathered in the final prewar years. In 1936, Alonzo Church, Emil Post, and Alan Turing published separate papers equating decision and computability procedures; in Berlin, Konrad Zuse began to build a universal calculator that used binary numbers, floating decimal point calculation, and the programming rules of Boolean logic. In 1937, Claude Shannon published a paper equating logic and circuity; Aiken and John Atanasoff worked out separate designs for calculating machines; George Stibitz built the first binary relay adder. In 1938, the Foxboro Company devised an electronic analog computer; Zuse completed a mechanical prototype of his hardware. In 1939, three seminal machines—Atanasoff's electronic calculator, Zuse's binary relay computer, and Stibitz's AT&T Model I—were all completed; IBM agreed to build Aiken's Mark I.

Even cybernetics, usually considered a postwar development, was largely anticipated in a paper published in 1940 by a British scientist, W. Ross Ashby. By the end of 1940, Stibitz had successfully demonstrated telecomputing and Atanasoff had begun conversations with John Mauchly that would help to shape ENIAC. Even though it would not appear for another six years, ENIAC remained less modern—in some respects—than the prewar machines: it used decimal rather than binary numbers and hence could not exploit Boolean logic, lacked a general-purpose central

8. Howard H. Aiken, "Proposed Automatic Calculating Machine," in *Origins of Digital Computers: Selected Papers,* 3rd ed., ed. Brian Randell (New York: Springer-Verlag, 1982), pp. 195-201.

9. John G. Brainerd, "Genesis of the ENIAC," *Technology and Culture,* 17(3):482-88 (1976).

10. Beniger, *Control Revolution,* chap. 9.

processing unit, and only partially distinguished processing from memory.

Twenty years before the first tube glowed in ENIAC, America's top four information-processing companies, with total revenues—in 1928 dollars—exceeding $150 million, were Remington Rand, National Cash Register, Burroughs, and IBM—names still recognizable as forebears of three of today's top five.[11] The companies owe their origins to four distinct innovations in the information machines of the 1870s and 1880s: the Remington typewriter, the cash register, the printing adder of William Burroughs, and Herman Hollerith's punch-card tabulating equipment. We must look here, to the technological and economic innovations of 1870-1900, and not to ENIAC and other developments of the 1940s, to find the truly revolutionary origins of the information society and of computing in science.

TELEMATICS AND SCIENCE

Today the most revolutionary impact of computing on science comes with the continuing proliferation—since the early 1970s—of microprocessing technology. Perhaps most important, at least in terms of its impact on science as a social system, has been the progressive convergence of information-processing and communications technologies—including mass media, telecommunications, and computers—in a single infrastructure of control.

A 1978 report commissioned by the president of France—an instant best-seller in that country and abroad—likened the growing interconnection of information-processing, communications, and control technologies throughout the world to an alteration in "the entire nervous system of social organization."[12] The same report introduced the neologism "telematics" for this most recent stage of the information society, although similar words had been suggested earlier, such as "compunications"—from "computing" and "communications"—by Anthony Oettinger and his colleagues at Harvard's Program on Information Resources Policy.[13]

Development of telematics in science also stems from the consolidation of the Control Revolution in computing in the late 1930s. By October 1939, George Stibitz, a Bell Laboratories physicist, had completed his Model I, a 450-relay calculating machine that input and output via teletype. The following September, at the annual meeting of the American Mathematical Society at Dartmouth College, Stibitz installed several teletype terminals and linked them—via Bell's long-distance system—to the Model I in Manhattan, some 200 miles away. Conference participants used this system to solve the complex-number equations for which the Model I had been hardwired—the first use of remote computing via telephone that would characterize the emergent "telematic society" thirty years later.[14]

Crucial to modern telematics is increasing digitalization, the coding into discontinuous values—usually two-valued or binary—of information varying

11. Stan Augarten, *Bit by Bit: An Illustrated History of Computers* (New York: Ticknor and Fields, 1984), p. 183.

12. Simon Nora and Alain Minc, *The Computerization of Society: A Report to the President of France* (Cambridge, MA: MIT Press, 1980), p. 3.

13. Anthony G. Oettinger, "Compunications in the National Decision-Making Process," in *Computers, Communications, and the Public Interest*, ed. Martin Greenberger (Baltimore, MD: Johns Hopkins University Press, 1971), pp. 73-114.

14. Evelyn Loveday, "George Stibitz and the Bell Labs Relay Computers," *Datamation*, 23(9):80-85 (1977).

continuously in time, whether a telephone conversation, a radio broadcast, or a television picture. Because most modern computers process digital information, the progressive digitalization of mass media and telecommunications content begins to blur earlier distinctions between the communication of information and its processing—as implied by the term "compunications"—as well as between people and machines. Digitalization makes communication from persons to machines, between machines, and even from machines to persons as easy as it is between persons. Also blurred are the distinctions between information types: numbers, words, pictures, and sounds, and eventually tastes, odors, and possibly even more complex sensations, all might one day be stored, processed, and communicated in the same digital form.

Through digitalization and telematics, currently scattered information—in diverse forms—will be progressively transformed into a generalized medium for processing and exchange by a global system, much as, centuries ago, the institution of common currencies and exchange rates began to transform local markets into a single world economy. We might expect the implications to be as profound for global science as the institution of money was for world trade. Indeed, digital electronic systems have already begun to replace money itself in many informational functions, only the most recent stage in a growing systemness of world society dating back at least to the Commercial Revolution of the fifteenth century.

SOCIETAL CHOICES: THE SOVIET CASE

Analogies between the integrative functions of digitalization and money seem particularly appropriate in the case of the Soviet Union, which may be as suspicious of telematics in its own society as it has been ideologically opposed to market control of its economy. Despite the fact that the Soviet computing industry covers the full range of products and, at least in this aspect, ranks behind only the United States and Japan, the Soviet Union still has not successfully mass-produced most of its computer devices.[15] Although "the Soviets have every capability to be as intellectually advanced as we are in computing," according to William McHenry, a Georgetown University specialist in Soviet and East European computing,[16] the country's computing infrastructure has developed slowly compared to those of other information societies, as has the range of Soviet applications.

This lag in telematic development is particularly ironic in historical perspective.[17] Soon after the Bolshevik revolution, Soviet economists Krassin and Grinko had already begun to develop the scientific theory of central economic planning and control based on continuous data collection, processing, and analysis. By the time Joseph Stalin had fully consolidated his leadership in 1929, Russia had the third largest amount of data-processing equipment—Hollerith card punches, sorters, and tabulators—in the world, behind only the United States and Germany. The Soviet "Gosplan" or five-year plans, according to an IBM-sponsored history of computing,

15. William McHenry, "Why Russian Computers Aren't Byting," *Times* (London), 28 July 1986.

16. W. David Gardner, "To Russia with Love: The Intrigue behind the Soviets' Illegal IBM Installations," *InformationWEEK*, 3 Feb. 1987, p. 41.

17. Beniger, *Control Revolution*, pp. 416-22.

"implied a level of control that could hardly have been attempted without machines for the rapid assessment of statistics."[18] By 1938, the Soviet mathematician L. V. Kantorovich had begun to develop the linear programming theory that remains central to military and industrial planning and control by computer in all industrial nations to this day.

With this early lead, and with a secure position as producer of computer hardware, why does the Soviet Union lag behind other advanced industrial nations in telematic development? At least until the rise of Mikhail Gorbachev, with his policies of *glasnost* ("openness") and *perestoyka* ("restructuring"), most Western explanations centered on the Soviet leadership's reluctance to allow greater access to information, with the decentralization of control that would result. This overlooks the probability that state control of information might still be maintained simply by controlling a few basics: paper, photocopy machinery and supplies, computer printers, ribbons, and other printing supplies. At Moscow's huge Lenin Library, for example, photocopying is limited to 2000 sheets a day for the entire facility, portable computers are banned as "copying devices," and admission permits have been revoked for attempting to bring a Western newspaper clipping into the building.[19]

A more likely explanation of the Soviet Union's lag in telematics would appear to be its lack of a considerable spectrum of possible computer applications—including communications, entertainment, education, and consumer goods—that foster development of hardware and software in other advanced industrial countries. Even after a twenty-year drive to introduce computer-based information systems into Soviet management, 92 percent of the country's approximately 44,000 industrial enterprises still send their data out to a Central Statistical Administration for processing. Organizations that do have their own computers use them almost entirely for batch processing and the production of periodic statistical reports, thereby limiting possibilities that telematics might be integrated with day-to-day activities.[20]

It might seem that the Soviet Union's military and surveillance infrastructure alone would produce sufficient demand for telematic development. Although computer use by the Soviet military and intelligence communities is indeed great, Seymour Goodman of the University of Arizona, currently studying the prospects of a Soviet information society, finds that this demand cannot alone provide an adequate material base for the larger societal transformation.[21]

Perhaps new Soviet leadership initiatives, such as the Interbranch Scientific Technological, a kind of interministry research and development collective, or the State Committee for Informatics—a variant on "telematics"—and Computer Technology, with greater control over the country's computer industry, will help to promote the country's telematic development. They will not do so, however, without difficulties arising from

18. Charles Eames and Ray Eames, *A Computer Perspective* (Cambridge, MA: Harvard University Press, 1973), p. 97.

19. Nicholas A. Ulanov, "Soviet Fear of the Knowledge Revolution," *Wall Street Journal,* 13 May 1986, p. 32.

20. William K. McHenry and Seymour E. Goodman, "MIS in Soviet Industrial Enterprises: The Limits of Reform from Above," *Communications of the ACM, 29*(11):1034-43 (Nov. 1986).

21. Seymour E. Goodman, *The Information Technologies in Soviet Society: Problems and Prospects* (Tucson: University of Arizona, Department of Management Information Systems, 1986).

the nature of control as a political as well as a technological problem. "Turning certain functions over to the computer presents a risk for the elite," McHenry writes of Soviet society, "the elite whose unique influence is based in part on controlling just a bit more information than subordinates do."[22]

The Soviet case shows that the information society does not spring spontaneously from advanced industrialization. Technological possibilities for control present societal choices, which are themselves subject to political control. It remains to be seen whether the Soviet Union can compete as a world economic leader without the frenetic development of microprocessing, computing, and telematics that seems increasingly to energize other information economies. It also remains to be seen whether Soviet scientists can maintain their preeminence, outside of telematics, as that technology increasingly transforms science itself into one world system.

CONTROL PROBLEMS IN SCIENCE

How important is science in the information society? In the classification scheme developed by Marc Porat for the Office of Telecommunications, U.S. Department of Commerce,[23] information workers are of three types: those whose output is an information product, constituting about 31 percent of U.S. information workers by compensation; those who process or move information within firms, constituting 64 percent; and those who operate information technology, who constitute 5 percent. Virtually all scientists can be found in the first category, which might be further divided into two subcategories: knowledge producers—including scientists—and knowledge distributors—educators, writers, editors, and the like.

Knowledge producers, about 19.0 percent of U.S. information workers by compensation, include scientific and technical workers—8.0 percent—and those who provide private information services like counseling and advising—11.0 percent. The former category includes not only workers in the sciences—2.4 percent—but also in engineering—5.4 percent. When we use the term "science," therefore, we mean roughly 2.0 to 8.0 percent—exclusive to inclusive—of all information workers by compensation. The importance to the information society of knowledge production—of which science and engineering constitute 40 percent—is obviously much greater.

Before digitalization and telematics transform science into a global system, science—much like the nineteenth-century railroads—will have to resolve certain problems of control. Other authors in this volume discuss potential barriers to the global systematization of science, including different languages—scientific and computer as well as natural—international politics, national information policies, laws concerning intellectual property and privacy, incompatibility of systems, and the continuing so-called arms race between code busters and encryptors.

As such barriers are overcome, science will face an even more challenging crisis of control, one brought about—ironically enough—by the very efforts intended to ease scientific communication. This crisis arises from the usually implicit

22. McHenry, "Why Russian Computers Aren't Byting."

23. Marc Uri Porat, *The Information Economy: Definition and Measurement* (Washington, DC: Department of Commerce, Office of Telecommunications, 1977), chap. 7.

assumption that science primarily consists—at its most macro level—of a one-way informational flow: knowledge is created, processed or refined, communicated and utilized, possibly to create still more knowledge. Informed by this model, much computerization of information systems slights the reciprocal or feedback signals by which scientific outputs are controlled. Such feedback, perhaps most familiar in the form of scientific citations but also as reputations of journals, editorial decisions, and a wide range of other such signals, does not represent knowledge produced but does confer status and authority differentially upon knowledge producers.

Exchange authority like that conferred by citation will be crucial to the control of any social system that has not been engineerined. It will be crucial in any interorganizational network not under extraorganizational control, for example, but not within any of the constituent organizations, where other more formal lines of authority prevail. Global science, lacking centralized control, must depend on countless loosely knit and continuously shifting networks of individual researchers—most of whom resist outside intervention—in communication that crisscrosses the borders of well over a hundred sovereign nations. Because these scientists work in a wide variety of universities and colleges, technical and trade schools, government and corporate laboratories, public and private research centers and foundations, and other institutions, science must span many organizational and professional boundaries as well.

If this did not present difficulties enough for control of science, science must also transcend political and institutional boundaries in a more problematic sense: for science to exist beyond the contributions of independently wealthy or self-motivated amateur savants—as indeed it did not exist before the last century—the pecuniary rewards of local institutions must somehow be made to depend on contributions to the larger system. How else might science secure the labor of legions of workers, not only to publish and otherwise disseminate information, but also to referee papers and proposals, review books, organize and chair professional meetings, serve as discussants, respond to queries and correspondence, and engage in a host of other relationships with scientists in other organizations and countries? How else might so vast an amount of unpaid labor be obtained to power the global system of science, were not its workers convinced that rewards from their own institutions depend—at least indirectly and in part—on more universal contributions?

Still more problematic, it is not enough merely to energize the system; it must also be motivated at a higher level, that is, controlled toward collective goals. Not only must component organizations reward each worker according to his or her status in the larger scientific community, but such status must differentially accrue to individuals in a way that will serve the global system. Obviously, statuses that characterize more traditional societies—such as statuses based on family, gender, and age—will not be optimal for science. Nor will such ascribed statuses be as responsive to rapid change as will achieved statuses like specialty and rank. Exchange authority as represented by scientific citation, in contrast, can not only respond to even the most rapid change, but communicates feedback about the value of each discrete output, from each of its potential users, for each of its potential applications, to the system as a whole. A computerization of science that fails to accommodate at least as finely tuned a control medium may experience system crisis—most acutely at the most global levels.

GENERALIZED MEDIA AND STATUS DIFFERENTIALS

Scientific citation not only provides direct feedback about the value of each individual contribution—paper, article, book, piece of correspondence, and so forth—but also serves directly to control total production—the system-level goal—through differential rewards to individuals. Not only are salary, promotion, and other organizational rewards for scientists influenced by citation, but so are less tangible but more universal rewards like standing in the scientific community. Exchange authority, as conferred through citation, serves to communicate status across institutional and cultural boundaries, where local organizational symbols—like formal titles and trappings of office—do not translate very well.

Thus citation serves science much as money—as noted by John Stuart Mill during the control crisis of the Industrial Revolution—can serve the political economy.[24] Both money and citation constitute symbolic systems that translate status across social contexts—the sense in which money is said to talk, or even to speak all languages. Like money, citation has several crucial characteristics of generalized symbolic media as first identified by Mill and elaborated by Talcott Parsons:[25] institutionalized function as a measure of value or status, no value in use but only in receipt or exchange, and non-zero-sum character in some contexts—just as money can be spent to create credit, citations can be made without loss to gain credit with those cited through norms of reciprocity.

Other social and control systems that span organizational and disciplinary or professional boundaries outside of science have been empirically shown to be integrated by generalized symbolic media of exchange. Among the educational, social service, legal, and health professions, for example, the concept of a professional referral is not only universally understood but plays a role analogous to that of citations in science.[26]

Higher-status scientists provide more information—they publish more and have wider networks—and receive more citations than do their lower-status colleagues. Analogously, in the four organizational sectors mentioned earlier, physicians and other professionals of higher status tend to supply specialized information in exchange for referrals, while their lower-status colleagues are more likely to serve as referral sources and information sinks in the same specialized system. In both the scientific and the other professional systems, in other words, movements of system information and referrals or citations communicate identical and opposite status relationships: information is better to give, while referrals or citations are better to receive, and these pairs of valued commodities tend to be exchanged directly

24. John Stuart Mill, *Principles of Political Economy, with Some of Their Applications to Social Philosophy,* 2 vols. (Boston: Little, Brown, 1848).

25. Talcott Parsons, "Social Structure and the Symbolic Media of Interchange," in *Approaches to the Study of Social Structure,* ed. Peter M. Blau (New York: Free Press, 1975), chap. 6, pp. 94-120.

26. James R. Beniger, *Trafficking in Drug Users: Professional Exchange Networks in the Control of Deviance* (New York: Cambridge University Press, 1983).

between pairs of individuals.

Most critical for control of both systems is a seemingly anomalous finding: the pattern of exchange associated with higher status in the larger interorganizational systems—information source, citation or referral sink—is precisely opposite of the pattern having higher status within any of the constituent organizations—referral or acknowledgment source, information sink. Within a leading scientist's own organization, that is, the scientist expects to receive valuable information from his or her research assistants, whom the scientist may acknowledge in a footnote; outside of the organization, however, the scientist expects to provide valuable information to many other scientists, who the scientist hopes will acknowledge him or her by citing his or her work.

Herein lies a commonplace irony of professional life: the same individuals who cannot find time to diffuse information and who shun the intra-institutional equivalent of referrals—whether of students, patients, clients, or the like—within their own organizations may devote considerable time to establishing themselves as informational sources—to enjoy the higher status that accrues through referral or citation—for anonymous individuals scattered throughout the world. Such individuals share many of the characteristics that Robert Merton first identified with cosmopolitans, in contrast to the locals, who narrowly focus their interests on local affairs.[27]

SCIENCE AS SYSTEM: ONE MARKET, MANY ORGANIZATIONS

Why do professionals divide into separate status groups with opposite patterns of exchange of system information and citations, acknowledgments, or referrals? This division, between what we might call organization-directed and system-directed scientists, following Riesman,[28] results from differences in the nature of control in formal or engineered systems, such as universities, research centers, and laboratories, versus free-market systems, such as global science in the ideal.

Formal organizational systems are designed—authority is distributed a priori—so that a few high-status individuals direct many others; they move the production of research or the processing of referrals and monitor organizational information for feedback to control these flows. To the extent that a market system lacks such design, by contrast, all producers control it in small measure. In science, researchers exploit information produced by others whose help they formally acknowledge in return, thereby distributing status and authority among a few leadership roles. Here the analogy between citations and money as market feedback to producers is straightforward.

Free-market patterns can obtain under all combinations of interpersonal or social network versus mass communication—in science, under all combinations of correspondence, draft, and preprint exchange versus formal publication. Whether in the networks of scientists called invisible colleges[29] or through a form of mass communication like an article in *Scientific American,* leading sources of system information will gain

27. Robert K. Merton, *Social Theory and Social Structure,* enlarged ed. (New York: Free Press, 1968), chap. 12, pp. 441-74.

28. David Riesman with Reuel Denny and Nathan Glazer, *The Lonely Crowd: A Study of the Changing American Character* (New Haven, CT: Yale University Press, 1950), chap. 1.

29. Diana Crane, *Invisible Colleges: Diffusion of Knowledge in Scientific Communities* (Chicago: University of Chicago Press, 1972).

status and authority—or celebrity, in the case of mass entertainment systems—whether these are invested by sociometric centrality, citations, or audience ratings.

It might seem that such systems, far from being free, are in fact controlled by the informational sources who enjoy highest system status. System-directed elite do impart direction to their subfields, but at the whim of their audiences of publishing researchers. Ultimate control is not only distributed, if not diffused, but also renegotiated day to day; compared to organizational control, it is much less centralized and never tenured. Continual renegotiation makes markets—under the control of exchange authority—adaptive even to rapid change.

Considering the opposite status implications of system information and citation giving and receiving, inside versus outside of formal organizations, it is hardly surprising that scientists divide into separate status groups having opposite patterns of exchange. A leading research methodologist, for example, will not need to solve problems for students or colleagues in order to maintain status achieved through publication and correspondence. A scientist whose research draws few citations, by contrast, might still hope to achieve high local status by advancement through the formal hierarchy in his or her own institution, where he or she will tend to monitor rather than to disseminate system information and to make acknowledgments and referrals rather than to receive them.

Because local reputations are not generalized symbolic media in the larger system, however, they do not translate well across organizational boundaries, except in partially generalized and partly symbolic media like letters of recommendation, and then only to middle and lower organizational levels. Status in the interorganizational or global system, by contrast, does translate well in most local contexts. This explains why organizational rewards tend to correspond to status on the global level, which has relatively few rewards of its own, Nobel Prizes providing a notable exception.

Translation from global system to organizational contexts remains imperfect, as, for example, when a scarce commodity like academic tenure is denied a talented scientist because she or he is a poor teacher or administrator. Fortunately for science, the generalized symbolic nature of global status assures that its failure to translate will occur less often than the reverse failure: when a poor scientist is promoted based on local contributions alone. Both types of failed translation bolster the view of science as a dual system, one in which the global goal to expand the universal body of scientific knowledge must continuously compete with more local but necessary functions ranging from teaching, fund raising, and administration to turning a profit and national defense. If the global goal were to cease to translate—via global status and authority—into organizational rewards, the system of science would deteriorate for lack of material incentives on the micro level and for want of control signals on the macro level.

POTENTIAL CRISIS OF CONTROL

Telematics threatens global science—as distinct from all the other work scientists do within their own organizations—with just such a crisis of control. Many involved with the computerization of information systems have predicted—some gleefully—a decline in the formal scientific paper, a blurring of the distinction between research notes and papers and between papers and the response to them by others, an increase in multiple authorship by scores or even hundreds who participate in a telematic discussion, and the decline of formal journals, editors, and the gatekeeping function more generally. Such developments, were they to occur, would destroy no less than the basic unit of research reporting, the paper; the organized system of communication, including conference presentations and journal publication; and the day-to-day feedback signals of exchange authority, like citation, that have served science since the Enlightenment.

Not everyone involved in the computerization of information systems intends to alter science so radically, of course. As Starr Roxanne Hiltz and Murray Turoff, who pioneered the Electronic Information Exchange System for Scientific Research Communities (EIES) in the mid-1970s, describe their own approach:

> The design philosophy is to start with the existing communication forms and functions of a user group and to build a system that accommodates or replicates such communication patterns at an overall increase in speed or efficiency and decrease in cost; and then add communication and information-processing capabilities not possible without the computer. The specific nature of any particular system optimized for human communication would vary somewhat from one organization and application to another.[30]

The difficulty here is in distinguishing those "communication forms and functions" that ought to be "accommodated[d] or replicate[d]" and those that ought to be speeded up, made more efficient or less expensive, or enhanced by computer, which might have unanticipated and possibly dysfunctional consequences like increasing volume, extending the user community, or weakening exchange authority. Who can know in advance whether speed, cost, efficiency, and other processing constraints—on a wide range of knowledge-creation, information-production, and communication activities—will be either crucial or incidental to any given social structure?

Even Hiltz and Turoff, based on their field trials of EIES, cite what they call "barriers and problems" in EIES's implementation in science. Reward systems are not currently constituted to measure and distribute credit for nonpublishing activities. Heavy capital investment in existing communication channels and delivery and distribution systems constitutes inertia inhibiting change. Competition for research funding undermines the sense of common interests necessary for free exchange. In subfields polarized by strong advocacy positions, many scientists do not want to confront each other except sporadically in print. Frequent communication among research sponsors, project leaders, and potential users may not benefit research.[31]

THE FUTURE

And what of the future? Most past attempts to forecast technological innovations or trends have proved worthless, at best, and certainly humbling. From the historical perspective outlined here, we might be tempted to conclude that—in at least broad outline—the Control Revolution will continue into the near future much as it has over the century just past.

30. Starr Roxanne Hiltz and Murray Turoff, *The Network Nation: Human Communication via Computer* (Reading, MA: Addison-Wesley, 1978), p. 19.

31. Ibid., pp. 249-50.

But even so cautious a forecast must be reconciled with perhaps the supreme dilemma of all control systems, including natural ecosystems: the greater the control, the more precarious, and the greater the potential for loss of control—with more disastrous consequences. The Chernobyl disaster and recent airline tragedies illustrate how much more precarious our control systems can make everyday life than it was in even the earliest days of industrialization.

Our information and control technologies continue to carry us, nevertheless, toward possible confrontations with the prospects of artificial intelligence, computer consciousness, even synthetic life. Intermediate forms already exist on both sides of life's boundaries: self-replicating polymers on the inorganic side, genetically engineered systems on the organic side. Considering the continued development and proliferation of nuclear weapons technology, under the control of worldwide and extraterrestrial systems of growing complexity, evolution's next stage might hinge on the question of which boundary we reach first: synthetic life, certainly one possible product of growing control, or the return of the planet to the inorganic level, another quite possible outcome—ironically enough—of the same revolution in control technology.

ANNALS, *AAPSS*, **495**, January 1988

Interdependence of Scientific and Technological Information and Its Relation to Public Decision Making

By MELVIN KRANZBERG

ABSTRACT: Recent breakthroughs in superconducting materials reveal the interdisciplinary thrust of modern science and technology. The past fragmentation and specialization of scientific and technical fields is countered by the need for joint research, occurring in several institutional contexts—university, governmental, and industrial—and on an international scale, all affecting the nature of science and technology communication. Recognition that the nation's security and economy depend upon its scientific and technological base means that government is increasingly concerned with its direction; at the same time, the internal dynamism of science and technology requires complex and costly endeavors, making them increasingly dependent upon governmental support and hence public approval. Informing the general public and policymakers adds external dimensions to communication needs. Furthermore, the complex nature of innovation requires greater and speedier communication within the science-technology community, met by more publications and advances in electronic communications, but not to the exclusion of the human element.

Melvin Kranzberg, Callaway Professor of the History of Science and Technology at Georgia Institute of Technology, was chief founder of the Society for the History of Technology and edited its quarterly journal, Technology and Culture, *for over two decades. In 1980-81 he was national president of Sigma Xi, the scientific research society. His many writings on scientific and technological history concentrate on various aspects of the interactions of technology with society and culture.*

ON 18 March 1987, an overflow audience attended the American Physical Society meeting in New York City to hear about revolutionary discoveries of rare-earth-based oxide ceramics that could perform as superconductors at relatively high temperatures. The meeting ignited the enthusiasm of scientists who foresaw application of these materials in virtually loss-free electric power transmission, super-strong magnets, magnetically levitated high-speed trains, and many electronic devices. Indeed, in a front-page story the next day, the *New York Times* headlined the meeting as the "Woodstock" of physics.

Until recently, superconductivity was an arcane field of research in quantum physics. But overcoming the temperature obstacles to superconductivity involved new kinds of ceramic materials and metals, so that researchers in ceramics, materials science, and engineering, electrical engineers, chemists, and workers in other fields contributed to this breakthrough. The fact that the new ceramic-based superconducting materials were being fabricated in many shapes and forms and were immediately being tested for possible applications provides evidence of the breakdown of the old dichotomy between science and technology.

PROFESSIONALIZATION AND SPECIALIZATION

Up until two centuries ago, the term "natural philosophy" covered all our modern sciences; only in the course of the nineteenth century did the separate scientific disciplines truly develop, manifested by increasing specialization in certain fields and by departmentalization within universities. Such specialization required high expertise and led to advances in knowledge along very narrow paths, just as the division of labor in other human activities also makes for greater efficiency and proficiency.[1]

The same specialization and professionalization occurred in technology. Known simply as the "arts" throughout most of history, not until the end of the eighteenth century did John Smeaton coin the term "civil engineering." As in the case of science, the nineteenth century brought the formation of specialized branches within the once monolithic field of civil engineering: mechanical, metallurgical and mining, electrical, chemical, and other varieties of engineering, corresponding to the growing complexity of technical problems and the entrance of new technical means.

This fragmentation of both scientific and technological knowledge into highly specialized compartments was reflected in the establishment of a host of specialized professional organizations, each publishing materials relating to its specific branch of knowledge. As the sciences and technologies became more specialized, they also began employing separate vocabularies and unique methodologies to solve problems peculiar to their fields. Furthermore, the torrent of knowledge in each particular field made it difficult to keep up with what was happening in a single specialized branch, let alone keep abreast of developments in other areas.

Other disadvantages of the growing specialization included the inability to communicate with researchers in other areas and to deal with problems that did not fit exactly into rigidly specialized fields. This latter became a serious handicap when new areas of science and technology opened up that required the

1. William J. Bouwsma et al., "Specialization and Professionalism within the University," *ACLS Newsletter*, 36(3-4):1-31 (Summer-Fall 1985).

unification of "seemingly disjoint specifics." Yet the consolidation of knowledge from several different fields enabled sharper analysis and synthesis and led to more efficient problem solving.[2] Such factors contributed to a contemporaneous development that ran counter to the growing specialization within scientific and technological fields, namely, the coming together of science and technology into a new relationship.

THE WEDDING OF SCIENCE AND TECHNOLOGY

The wedding of science and technology is a relatively recent phenomenon.[3] Science historically stems from philosophical and speculative inquiries about nature in the ancient societies that fathered our Western tradition. Technology, even more ancient, stems from the arts and crafts carried on by slaves and other humble folk who had little contact with the philosophical tradition.

This dichotomy persisted in the Middle Ages. In that Christian age, study of the physical universe was regarded as an adjunct to the understanding of God, and science was a branch of theology. Technology, still carried on in the craft tradition by servile beings, remained wholly separate from investigations of the meaning of the physical universe.

By the seventeenth century, however, a series of social and economic changes produced a secular attitude that freed science from its preoccupation with theological questions, and the growing level of wealth stimulated the consumption of material goods and hence spurred the advance of technology. Yet the association between science and technology remained both tentative and ambivalent.

Although it cannot be said that the Scientific Revolution was impelled by technical considerations, the rationale for scientific investigation began emphasizing its utility for humanity. Nowhere is this better seen than in the writings of Sir Francis Bacon, who claimed that scientific knowledge was especially powerful in making nature useful to humanity and thus improving the human condition. Bacon's followers founded the Royal Society, whose principle of "promoting useful knowledge" was embraced by the American Philosophical Society, founded by Benjamin Franklin in Philadelphia in 1743.

Nevertheless, the Royal Society and its Continental counterparts became devoted almost entirely to basic science rather than utilitarian applications while the landmark inventions of the Industrial Revolution relied little, if at all, upon scientific findings and came from men who were tinkerers without special training in science. Yet the revolutionary technical developments, especially the steam engine, forced scientists to think about the relations between heat, force, energy, and work performed, leading to the science of thermodynamics and prompting the observation that thermodynamics owes more to the steam engine than the steam engine owes to thermodynamics.

Technology as applied science

Early in the nineteenth century, Faraday's work on electromagnetism and the

2. Arnold Reisman, "Expansion of Knowledge via Consolidation of Knowledge" (Technical Memorandum #594, Department of Operations Research, Case Western Reserve University, 1986).

3. Melvin Kranzberg, "The Wedding of Science and Technology: A Very Modern Marriage," in *Technology and Science: Important Distinctions for Liberal Arts Colleges,* ed. J. H. Burnett (Davidson, NC: Davidson College, 1984), pp. 27-37.

birth of the electrical industry helped give rise to the notion that technology was merely the application of scientific discoveries. Technology itself was becoming more scientific, however, in the sense that new technologies, founded upon recent research in electromagnetism and chemistry, could not be fully used until the engineers learned the scientific principles involved. And because the scientists had not investigated these phenomena in terms of application, the technologists had to conduct their own scientific investigations in order to fulfill their goals. One scholar described that technological development as the "mirror-image twin" of science.[4]

But just because technology was developing a science base does not mean that scientists and technologists were coming together professionally or educationally. Instead, engineers organized their own professional societies and issued their own publications. Furthermore, many institutions of higher learning did not admit engineering to their curricula. This exclusion fostered the founding of separate institutions for higher education in technology during the last half of the nineteenth century, especially when the Morrill Act led to the establishment of land-grant universities and of "agricultural and mechanical" institutions.

When the industrial research laboratory—the research and development lab, which had proved so successful in Germany—was introduced into the United States, it finally provided an institutional mechanism for uniting science and technology. Instead of scientists and engineers working apart in university laboratories or singly in their attics and cellars, industrial research brought together scientists and engineers from various fields in an organizational context that had as its aim the development of utilizable products and, if need be, the undertaking of basic scientific research in order to attain that goal.

Technological instrumentation for scientific advance

While engineers were becoming increasingly involved in work that had previously been the exclusive domain of science, scientists in their turn began requiring more technical expertise for their experimental investigations. Indeed, Derek Price found that all progress in experimental physics in the early decades of this century seemed to depend upon very ingenious craftsmen "with brains in their fingertips," and he implied that modern science is really "applied technology."[5] Similarly, Yakov M. Rabkin has shown how the adoption of infrared spectroscopy "significantly expanded the limits of organic and inorganic chemical research and enabled chemists to tackle problems whose scope and nature had hitherto eluded fruitful investigation."[6]

Phillip H. Abelson states that today technical instruments "often shape the conduct of research. They make discoveries possible. They determine what discoveries will be made."[7] As a result, technology can no longer be regarded as

4. Edwin T. Layton, "Mirror-Image Twins: The Communities of Science and Technology in 19th-Century America," *Technology and Culture*, 12:562-80 (1971).

5. Derek de Solla Price, "Of Sealing Wax and String," *Natural History*, 93(1):49-56 (1984).

6. Yakov M. Rabkin, "Technological Innovation in Science: The Adoption of Infrared Spectroscopy by Chemists," *ISIS*, 78(291):31 (Mar. 1987).

7. Phillip H. Abelson, "Instrumentation and Computers," *American Scientist*, 74:182 (Mar.-Apr. 1986).

a "dependent variable, drawing its ideas from and parasitic upon science; rather it is an equal partner contributing at least as much to the common stock as it draws out."[8]

Fragmentation yet togetherness

The contradictory tendencies toward fragmented specialization within science and technology along with the need for togetherness are displayed in scientific and engineering education. During the nineteenth and early twentieth centuries, academic disciplines in the sciences—for example, physics, chemistry, biology—emerged as the dominant unit of knowledge production, providing disciplinary coherence and a source of control over research priorities and reputations.[9] Engineering education underwent the same disciplinary specialization, as new technical fields emerged with the advance of industrialization.[10]

In the past few decades, the internal dynamism of science and technology has changed the nature and scope of both scientific and technological activities, and the specialized disciplinary distinctions have begun to fade. The reason is that the boundaries of physical nature and the complexities of modern technical devices do not correspond to the disciplinary boundaries of the separate fields as they developed in the nineteenth century. Synthesis of two or more fields proved necessary when it was learned that certain problems in biology had, for example, chemical and physical parameters; that materials had chemical, physical, and electrical properties; and the like. The result was a host of new scientific interdisciplines, such as biochemistry, astrophysics, and geophysics.

In the past two decades, this blurring of traditional fields has also been stimulated by social and political pressures upon both science and technology. Thus growing concern about the environment in the 1960s led to the development of environmental engineering and ecology, and the energy crisis of the 1970s aroused interest in energy propagation and devices that transcended the traditional branches of science and engineering.

The result was a proliferation of scientific and engineering subdisciplines, along with a growing need for collaboration between science and technology. With scientists and engineers working together on the same or related projects, interdisciplinary communication became essential to advances in research.

INSTITUTIONAL RESPONSES TO INFORMATIONAL NEEDS

The need to exchange information both within and between scientific and technological disciplines had a profound impact upon their organizational structure. Within universities, rather than disturb the feudal principalities of disciplinary departments that had grown up in the preceding century, special institutional mechanisms were created to take care of the new hybrid sciences and engineering fields. These usually took the form of centers or institutes, which combined various subbranches of sciences and engineering for research on common problems.

8. Donald S.L. Cardwell, *Turning Points in Western Technology* (New York: Neale Watson, 1972), p. ix.

9. R. Whitley, *The Intellectual and Social Organizations of the Sciences* (New York: Oxford University Press, Clarendon Press, 1984).

10. Melvin Kranzberg, ed., *Technological Education—Technological Style* (San Francisco: San Francisco Press, 1986), pp. 1-13, 75-84.

Subgroups were formed within professional associations to take account of the new specialists and specialities, but in most cases the older professional societies expanded in order to include the new hybrid fields. Thus the American Chemical Society and the Institute of Electrical and Electronics Engineers developed many branches and suborganizations within the larger parent society, each with its own publications. For example, the Engineering Management Society functioned organizationally within the larger Institute of Electrical and Electronics Engineers' framework, focusing on the "broad spectrum of functional areas" involved in the management of research of all kinds.[11]

Such institutional changes seek to bring together the many different sciences and technical fields involved in areas of research. Meetings and publications serve as mechanisms for the communication of ideas across previously fragmented and specialized fields and across researchers with differing institutional affiliations, and, in a sense, provide for the unification of knowledge. The breakthroughs in superconductivity reported at the meeting of the American Physical Society in mid-March 1987 acted as an immediate and powerful stimulus to further research in that field. Within the next six weeks, additional major advances were reported at the annual meetings of the Materials Research Society and the American Ceramic Society. These reports on superconductivity came from university, industrial, and national laboratories—and all of them were excitedly exchanging information.

11. Dundar F. Kocaoglu, "Editorial: Engineering Management—New Perspectives," *IEEE Transactions on Engineering Management,* EM-33, 1:1 (Feb. 1986).

Academic-governmental-industrial research collaboration

World War II, with its emphasis upon electronic, nuclear, and medical research, produced new and powerful national laboratories devoted to scientific and technological goals. Furthermore, some universities established research centers under governmental auspices, and government support of scientific research was extended further after the war with the establishment of the National Science Foundation. The government emerged as the major sponsor of scientific and technological research in the universities, and academic scientists and engineers remained somewhat aloof from business corporations.

But then came several developments that narrowed the traditional gulf between academic science and industrial activities. Most important was the discovery in genetics that molecules of deoxyribonucleic acid could be cut, spliced, and synthesized in many different ways to produce enzymes, growth hormones, and inhibiting factors. It was almost immediately recognized that these would allow for many commercial applications and, not surprisingly, the university as a corporate entity and university scientists as individuals began entering into research arrangements with biotechnology companies.

Along similar lines has been the development of another institutional mechanism—namely, a consortium of American corporations. The Microelectronics and Computer Technology Corporation was established in 1983 to pool research efforts that would lead to innovations to be shared by all. This consortium and the cooperative biotechnology activities of universities and industry thus represent new institutional arrangements to facilitate the communication of information essential for innovation.

International communication

Linkages between academia, government, and industrial scientific and technological research are by no means confined to the United States. Indeed, much of Japan's success in developing its electronics industry is attributed to the government's role in encouraging the research efforts of individual corporations. Similarly, the National University of Singapore has set up an Innovation Centre to promote interaction between engineering faculty and industry.[12] In Britain, where university faculty traditionally have stayed aloof from industrial activities, new types of industry-university collaboration are now being explored.[13]

But this has resulted in a paradoxical situation. On the one hand, scientists throughout the world share the tradition of communicating the results of their research to scientists everywhere. This has been furthered by the development of computer networks supporting electronic mail and other forms of efficient exchange of information, and these networks have expanded rapidly around the world. On the other hand, the value of the technology that emerges is so important to a nation's economy, its military strength, and even cultural chauvinism that governments seek to control the flow of scientific and technical information. Thus the new collaboration between government, industry, and universities raises fears that the traditional open dissemination of research results might be hampered by industry's preoccupation with proprietary information and by governmental concerns with national security.

12. *Science & Technology Quarterly* (Science Council of Singapore), 6(2):6, 12-14 (Jan. 1987).

13. Kevin A. Boyle, "Technology Transfer between Universities and the U.K. Offshore Industry," *IEEE Transactions on Engineering Management,* EM-33, 1:33-42 (Feb. 1986).

SCIENCE AND THE PUBLIC

Following World War II, both scientists and engineers were smugly confident of public approval because of the triumphs of modern science and technology. In the 1960s and 1970s, however, the political innocence of the scientific and technological communities began to fade when there arose widespread questioning of the benefits produced by science and technology, and there was fear that public support would diminish. As a result, these communities, which had hitherto communicated only within themselves and then gradually with one another, began being concerned about communicating with the outside world in ways that might influence public policies affecting their work. Hence the National Academy of Sciences established a Committee on Science and Public Policy, the American Association for the Advancement of Science followed suit by creating a Committee on Science, Engineering, and Public Policy and a Committee on Public Understanding of Science, and the National Science Foundation established a Division of Policy Research and Analysis. In response to this new interest in a field that had heretofore been nonexistent, universities initiated programs that studied the various parameters of science policy.

Scientific publications, such as *Scientific American,* began to report regularly upon sociopolitical aspects of the scientific endeavor, and the American Association for the Advancement of Science developed a group of science journalists for accurate reporting on governmental matters affecting scientific and technological undertakings. In 1984 the National Academy of Sciences, the National Academy of Engineering, and the Institute of Medicine launched a journal, *Issues in Science and Technology,* to present informed opinions regarding public policy issues affecting science and technology.

A number of factors complicate the efforts of the scientific community to communicate with the public. For one thing, the community is often at odds with itself, and members are frequently in competition for research funding, resulting in fears that some new scientific discovery in one field might diminish funding in other fields. For another, public decisions regarding science and technology matters are also influenced by traditional pork-barrel politics. Some institutions have gone directly to Congress to obtain funding for research facilities without going through the usual peer-review procedures of governmental science agencies—to the dismay of the Association of American Universities.[14] Finally, as Edward Wenk has shown, public policy on issues involving a choice between different scientific applications can be influenced by a sudden crisis or pressures exerted by various political factions.[15]

COMMUNICATIONS WITHIN SCIENCE-TECHNOLOGY

Traditionally, the principal means of communication between scientists was the printed word—in books, scientific journals, and the proceedings of meetings. The same holds true for technology, although the proprietary nature of much technical information provided a contradictory manifestation, namely, the desire to maintain trade secrecy against the desire to profit through patent registration, which involves publication. But in both science and technology the impulse was to lay claim to a discovery or innovation by being the first to publish results or by obtaining a patent.

In the twentieth century the growth in the number of scientists and engineers and the increasing number of specialized fields and professional organizations led to an avalanche of publications catering to these specialized needs. By the beginning of the 1980s, some 8000 journals were being published in biology and medicine alone—and the same phenomenon was repeated in other fields.[16]

Despite the vast proliferation of scientific and technical publications, both scientists and engineers complained of delays of a year or more in publishing their findings. Innovations, such as word-processing machines and computerized photocomposition, promised to make the publication process speedier and more economical. But prices kept going up because the journals were growing in size, reflecting the increasing amount of research. Thus the *Journal of Physical Chemistry* went from about 5000 pages in 1982 to 7000 pages in 1986; *Organic and Metallic Chemistry* grew from 1800 to 2600 pages. Widespread dissatisfaction among researchers persisted in regard to the "leaping prices" of the publications and the time lag in publication, the latter caused partially by the time-consuming, yet essential, refereeing process.[17]

14. *Science,* 22 May 1987, p. 909.

15. Edward Wenk, *Tradeoffs* (Baltimore, MD: Johns Hopkins University Press, 1986).

16. Fitz Machlup and Kenneth Leeson, *Information through the Printed Word: The Dissemination of Scholarly, Scientific, and Intellectual Knowledge,* 4 vols. (New York: Praeger, 1978-80).

17. Constance Holden, "Libraries Stunned by

The problem of delay in publication was brought to the fore by the explosion in superconducting research at the beginning of 1987. As noted earlier in this article, a once fairly obscure field of physical theory and materials research became highly visible, and researchers immediately began experiments in that area and then rushed to submit reports to the journals. Why were they so eager to publish their experimental results so quickly? Developments were occurring so rapidly in this field that the researchers feared that their discoveries would become obsolete before they had even been reported in the literature, and, perhaps more important, they desired to achieve recognition for their research results before others working in the same area. Hence those journals in the fields represented by the new superconducting materials research, especially the *Physical Review* and its offspring, were overwhelmed with reports awaiting publication.

Of course, scientific publications serve a greater purpose than satisfying the ego demands of first discovery; they are the primary means for scientists to keep abreast of their fields. In the exciting superconductivity race, the scientists were too impatient to await journal publication, so they turned to newspaper accounts and press releases to keep up to date. But because so much was happening so quickly in this field, some scientists exchanged daily telephone conversations or electronic mail with distant colleagues who were engaged in research paralleling theirs.[18] In brief, new electronic means of communication began supplementing the older system of communicating research through publication.

Scientists increasingly are engaging in networking, that is, in using computers as mechanisms for the exchange of data. Computers allow instant access to several categories of databases, including those containing: (1) full texts; (2) references, listing bibliographical items with descriptions of their contents; and (3) source data bases, containing facts available in handbooks, directories, reference guides, and statistical reports of many different varieties; for example, the fact data base of the American Chemical Society has information on the structures and properties of some 7 million compounds. Because the amount of information stored in both bibliographical and factual data bases is growing at a rapid pace, "computer retrieval of information is the answer to what would otherwise be a crippling proliferation of scientific literature."[19]

These on-line data bases are becoming increasingly specialized and enable scholars to locate sources of information on many different subjects and to obtain material that might otherwise be extremely difficult to collect. Thus, for example, the major academic libraries in the United States and Canada share bibliographical information through an on-line Research Libraries Information Network. Furthermore, computer systems that are interconnecting on a global basis enable scientists not only to tap large numbers of data bases but also to communicate with other people in the same field, exchange research data, and work out problems. Inasmuch as information exchange is a vital part of the

Journal Price Increases," *Science,* 22 May 1987, pp. 908-9.

18. Kim McDonald, "Race to Produce Superconducting Materials Stirs Concern over Publications and Patents," *Chronicle of Higher Education,* 29 Apr. 1987.

19. Phillip H. Abelson, "Instrumentation and Computers," *American Scientist,* 74:182 (Mar.-Apr. 1986).

scientific and engineering professions, these electronic capabilities greatly enlarge the possibility for scientific and technical advance.[20]

Because the new communication technologies affect the way that information is created, distributed, and used, however, they create new problems.[21] For example, how can existing copyright law, based on the originality of works in individual workmanship, be administered when the work involves many authors, worldwide collaboration, and dynamically changing materials? How can it be enforced when vast amounts of copyrighted materials can be copied, reprocessed, and traded without the knowledge or permission of the copyright holders? Does the information in a process, for example, computer software, represent a new kind of intellectual property? Could or should the original copyright holder control the repackaging of information—derivative use—made possible by the new technologies? Such questions would seem to indicate that the old copyright and patent laws have been made obsolete and hence must be overhauled to take account of the new electronic communications capabilities.

While computerized data bases and electronic communications allow instantaneous access to the growing flood of research, thereby casting into doubt the bases of copyright and patent laws, however, they also present problems of their own. For example, how does one find the needle in the mountainous haystack of software? How does one know what information is worthwhile and what is not? Traditional publication means allowed for some peer review to weed out information that was irrelevant, biased, or just plain wrong. Although such evaluation was sometimes subject to shortcomings, it nevertheless performed a service that is absent from the new means of publication via electronic networks. Just because the haystack is growing larger does not necessarily mean that more needles are implanted within it. Also, the lack of repeated proofreading increases the possibility of errors—without the requisite errata sheet.

In addition, the fact that more information is available in rapid form does not mean that the computerized data bases will be scanned and searched by people from other disciplines who might profit from this cascade of information. The old problem of keeping abreast of scientific and technological developments in fields other than one's own has not diminished. Informally sampling my colleagues on the faculty of the Georgia Institute of Technology, I discovered that while they scrupulously followed the specialized reports, both published and computerized, in their own fields, they still obtained generalized information regarding other fields by reading journals such as *Scientific American, American Scientist,* and *Science.*

Increased science reporting in the newspapers has added a new element. Inasmuch as the newspapers provide immediate, brief, and readable accounts of the latest developments, these, too, have become must reading. In brief, science journalism not only brings science news to the general public but also serves as a means of conveying that

20. John S. Mayo, "The Evolution of Information Technologies," in *Information Technologies and Social Transformation,* ed. Bruce R. Guiley (Washington, DC: National Academy Press, 1985), pp. 7-33.

21. U.S., Congress, Office of Technology Assessment, *Intellectual Property Rights in an Age of Electronics and Information,* 052-003-01036-4 (Washington, DC: Government Printing Office, 1986).

information to scientists and engineers outside the fields that are being reported. Not surprisingly, I also discovered in my informal survey that many scientists find out about what was happening in other laboratories in their own institutions only by reading about it in the newspaper!

While scientists recognize that there might be some omissions or distortions in newspaper accounts, the accounts at least provide them with some immediate awareness of interesting work in other fields. If any of these seem germane to their own research, they seek further information by asking for more details from their colleagues who specialize in those fields.

So again we come to the basic element in the communication of scientific and technical information: the human being!

THE HUMAN COMMUNICATION LINK

The human role in the communication of scientific and technological information was brought to the attention of students of the innovation process almost two decades ago by Thomas J. Allen, who first used the term "gatekeeper" to describe those individuals in research laboratories who serve as sources for the transfer of technical information from outside the organization to their research colleagues.[22] These gatekeepers had more exposure to the literature and more frequent contacts outside the organization, had been extremely active in research themselves, and provided technical information contacts outside the research and development laboratory to their colleagues. Despite changes in the organizational structure and the dynamics of today's research environment, recent studies indicate that gatekeepers remain a major source for the transmission of scientific and technological information.[23]

In the making of new discoveries and innovations, the human link in transmitting key information is especially important. The personal accounts of scientists and engineers whose research has achieved the status of being classic reveal how scientists and engineers derive stimulation and inspiration from exchanging ideas with one another; the combined knowledge of different individuals bouncing their ideas off one another achieves synergistic results.[24] Their explanations of how they arrived at their important results reveal the human nature of the scientific and technological enterprise.

This same immediate and personal quality in the transmission of scientific and technical information was displayed in the superconductivity Woodstock meeting in March 1987 with which I began this account. The scientists who crowded into the superconductivity session at the American Physical Society meeting knew that they could eventually read the reports in published proceedings and accounts. Yet they arrived early and stayed late, showing once again that the scientific and technological enterprises cater to very human needs and activities.

22. Thomas J. Allen and Stephen I. Cohen, "Information Flow in Research and Development Laboratories," *Administrative Science Quarterly,* 14:12-19 (1969).

23. Robert L. Taylor, "The Impact of Organizational Change on the Technological Gatekeeper Role," *IEEE Transactions on Engineering Management,* EM-33, 1:12-16 (Feb. 1986).

24. Arnold Thackray, comp., *Contemporary Classics in Engineering and Applied Science* (Philadelphia: ISI Press, 1986).

ANNALS, *AAPSS*, **495**, January 1988

Vanishing Intellectual Boundaries: Virtual Networking and the Loss of Sovereignty and Control

By RICHARD JAY SOLOMON

ABSTRACT: The programmable electronic computer is unique in the history of technology. It is a machine that may change its own instructions and can be programmed to emulate or simulate and thereby become any other machine. This definition of the stored-program computer is often misconceived by analysts who focus on narrow issues of what the machine does with information flow, instead of how it changes what flows. Because of this machine's capabilities, and that of ancillary electronics and optics technologies, over the next two decades we will probably witness a watershed in industrialization and the practice of science. The new technologies are also likely to create problems for international trade, the specialization of labor, and the use of national resources: they will make it virtually impossible to protect intellectual creations, including creations that underpin the workings of computer infrastructure; they will muddle definitions of legal, political, social, and economic boundaries—and impinge on paradigms of national and perhaps personal and corporate sovereignty; and they will accelerate the decentralization of decision making and permit stronger remote management.

Richard Jay Solomon is a research associate in the Massachusetts Institute of Technology's Research Program on Communications Policy, a consultant at the MIT Media Laboratory, and editor of International Networks, *a newsletter on telecommunications and computer technology and international policy. He is currently a consultant on telecommunications and computers for the Organization for Economic Cooperation and Development, Paris, and a columnist for* Telecommunications Magazine. *He has authored numerous papers dealing with the impacts of the computer and digital telecommunications.*

DURING the initial era of expensive data-processing machines, computers were run in a dominant batch-processing mode, occasionally connected to other computers over definable links that could be monitored by humans. This mode was followed by access to the central processing unit (CPU) by means of time-sharing, in what was essentially a series of batch processes until interactive programs, such as programs for word processing and spreadsheets, were developed. These two modes were an aberration in the application of the technology, though the time-sharing architecture led directly to the operating systems used on today's dispersed microcomputer workstations.

Even the initial spread of packet-switched data networks was designed with the concepts of centrally maintained mainframes remotely extended to user terminals, for remote job entry or time-sharing. But, instead of mainframes run by data-processing departments and data-processing specialists, information processing is rapidly shifting to a global set of interconnected, multiple virtual networks of inexpensive machines that respect no boundaries. The CPU now is smaller than a thumbnail, and most computer workstations actually house several CPUs. Dispersing the computer power is simply more efficient and can allow functions that are prohibitive on expensive mainframes. But dispersion does not imply independence; the stand-alone word processor or unlinked microcomputer will, over time, also be seen as an aberration.

GLOBAL BROADBAND TRENDS

The trend is inescapable. The level of complexity of this new and quite spontaneous infrastructure is becoming so vast that any attempt to dismantle merely portions of its interlocking elements even at the current stage of evolution would expose industrial society to immense disruption. Complexity is both the new infrastructure's strength in terms of redundancy, and its weakness in terms of a vulnerable, socioeconomic Achilles' heel.

LANs and B-ISDN

Examples of the radical shift in architecture, and hence the shift in data-processing paradigms, are the broadband local area network (LAN) and proposed broadband integrated services digital networks (B-ISDN). With LANs, multiple CPUs perform a range of functions from driving interactive terminals to running background processes, serving as telecommunications switching devices, and printing. LANs can easily extend across a metropolitan region via private circuits on public telecommunications carriers, so as to access data and control remote processes and processors.

In the past few years, a number of efforts have begun to design interregional public broadband networks, which would provide comprehensive, switched, gigabit per second service as early as 1992. These began as integrated services digital networks and now go under the name of "B-ISDN." Most of the designs appear to be converging on a very fast but novel implementation of today's packet networks, which would be capable of carrying data, voice, videophone, and even high-definition television.[1]

1. For a more technical discussion of the shift to B-ISDN, see Richard Jay Solomon, "Open Network Architectures and Broadband ISDN: The Joker in the Regulatory Deck," in the *Proceedings* of the ICCC-ISDN 1987 Conference (International Council for Computer Communica-

B-ISDN will be preceded by a narrowband upgrading of existing telephone operations, with an increase of an order of magnitude in local telephone circuit capacity and with a change of several orders of magnitude in the speed of a connection. This gives us a glimpse into a future of telecommunications technology that permits entering to and extracting from a CPU's local memory space huge amounts of data within microsecond time frames and that allows access to global data bases. One result will be a shared memory of almost unlimited size distributed anywhere—and therefore everywhere. Another will be the spawning of novel network architectures and new types of global networks that will change the infrastructure of both science and industry.

The microcomputer connection

With the advent of the 32-bit processor and matching bus, microcomputers have now reached the mainframe-on-a-desk stage. Some micros linked together in LANs can outperform some mainframe architectures. Such powerful, and dispersed, micros are able to take transparent control of other remote machines via direct access to a CPU's local random-access memory space, accomplishing a significant step toward B-ISDN application. Computers require access speeds ranging in the microseconds to permit such memory applications—something a broadband link can provide only terrestrially, obviating a technology that is itself less than 20 years old: satellites.

A shift in boundaries

The interconnection of computers and telecommunications technology at the level of the CPU and its associated memory is not a trivial change. By extending the computer's memory across telecommunications networks, a virtual network is formed—a network with logical, rather than physical, boundaries. With virtual networks, the network itself is transparent to the user, the user more likely being another CPU chip rather than a human.

Control passes to the program or process that interconnects the devices. This control may belong to the user program as much as to the telecommunications carrier's program. To fix boundaries in virtual networks becomes an exercise in futility because there is no fixed quantity at a fixed place in a fixed time that is measurable.

The inherent designs of the stored-program computer make it impossible to determine what data are where or when, nor will there be any incentive on the part of the user or programmer to even care about such matters. A process takes place within the CPU and may extend to the outer reaches of its connections, which could be halfway around the world. A process is a series of states for fleeting intervals. Unlike telephone or telegraph connections, on a digital network the electrical or optical connections are not what count because the digital information is transmitted either by synchronized timing or by asynchronous but coordinated packets. Another way to look at a virtual digital network is that the signals are located in time, not space.

tions, Dallas, TX, 15-17 Sept. 1987); Richard Jay Solomon and Loretta Anania, "Open Networks—The Beauty and the Beast: Virtual Networking in B-ISDN," *Telecommunications* (North American ed.) (Sept. 1987); R[ichard] J[ay] S[olomon], "The Shift to Broadband ISDN," *International Networks Newsletter*, 15 Sept. 1987; and Richard Jay Solomon, "Changing the Nature of Telecommunications Networks," *Intermedia*, 14(3):30-33 (May 1986).

The failure of regulation

Virtual networks overwhelm any attempt to control the loci of the data flow. The spread of packet-switching nets in the 1970s brought a reaction from governments to control transborder data flows for a variety of reasons, ranging from personal privacy to violations of telephone tariffs. But the growth of the nets has made a mockery of attempts at regulation, because there is simply no way to know what the networks are carrying, much less how and where the data are moving, without getting inside of a user's complex, high-level application program, not to mention the user's organization. Virtual networks are transparent to governments as well as to users, and the dismantling of national sovereignty is traceable to the level of the computer's basic input-output system.

THE COMPUTER

Taking the concept of a machine liberally, we can conceive that any process that can be expressed in a series of steps, or mathematical algorithms, can be emulated by a programmable digital computer.[2] Indeed, the correct definition of a stored-program process indicates that the process is the machine, as expressed by its functioning within the digital computer. Because the parameters and functions of this novel machine can be modified by modifying the program, software is much more powerful than would be implied by the mere characterization of a program as a series of machine instructions.

As a result of other twentieth-century technology, computers have become permanently inexpensive while CPUs have become very small and have taken on the character of fundamental commodities, like wheat or machine screws. This has caused a shift in computer application as well as computer paradigms. The cost of microcomputer CPUs and components has decreased as their power has accelerated; this relationship has come about faster than equivalent ratios for mainframes and minicomputers. The crossover point may be within the next few years, when the concepts of microcomputers, minicomputers, and mainframes are expected to merge. At that time, the microcomputer complex called workstation, now pervading the laboratories of most fields of science and engineering, is likely to become a supercomputer-on-a-desk, interconnected as a matter of course with various other distributed digital machines via a series of global networks.

Machine intelligence

Using a number of clever arithmetical techniques and relatively conventional sensor and transducer technologies, the programmable processor can be made to store, re-create, modify, and analyze virtually anything that a human can hear, see, touch, smell, and taste—and quite a lot that humans cannot sense! The machine does these things by accepting analogues of external signals and converting these signals to digital form using preprogrammed rules. An example of how digital devices can interface with and manipulate the human world of hearing are the computers that digitize sound, a process called pulse code modulation.[3] These computers were

2. A good description for the layperson on how a digital stored program computer works is in Joseph Weizenbaum, *Computer Power and Human Reason* (New York: Freeman, 1976), chaps. 2 and 3. See also Tom Forester, ed., *The Microelectronics Revolution* (Cambridge: MIT Press, 1981).

3. Colin Cherry, *On Human Communication*, 2d ed. (Cambridge: MIT Press, 1981).

among the first rudimentary expert systems but now have evolved into full-scale stored-program machines for switching and sound enhancement. Extension of this concept to visual systems permits image enhancement and computer animation and in the coming years will greatly improve television systems through manipulation, compression, and anticipation of an image's motion components.[4]

The digital communication lines that can mix voice, pictures, and data indiscriminately are all based on variations of this technology. This is the new dominant mode of all telecommunications installations worldwide—a powerful example of the merging of technologies and the blurring of lines of control by the carriers of communications.

Computer data communications

By its very nature, an electronic computer must communicate with other computerlike devices and must communicate using its internal mathematical code consisting of binary digits, or bits. These other devices may be peripherals essential for its stored-program operation, such as temporary memory buffers normally physically adjacent to, or incorporated as part of, the central processor; or communications may be with external memory devices, sensors, transducers, and other digital computers, which may be at a great distance from the processor. Within the processing environment, distance is meaningless, as long as the signals reach their destination within an allotted time period and without any degradation whatever.

4. For a bibliography, see Richard Jay Solomon, "Intellectual Property and the New Computer-Based Media" (Report to the Office of Technology Assessment, U.S. Congress, Aug. 1984).

Perfect information transfer in the form of digital data is another corollary to the information age: all digital data must be captured and re-created in a form that is not worse—and if possible better—than the original.

Standards

The making of standards has immense implications for the control of information related to scientific research, national economic goals, and market power. It is also of crucial importance to the establishment of an open philosophy toward computer communications. Digital encoding used by computers has many variants, and the application of sophisticated codes has created numerous opportunities to improve imperfect communication channels. Encoding can also be used, however, to establish market barriers by manufacturers intent on controlling trade secrets or by scientific groups who want to limit access to their data.

Herein lies another paradox, for one can limit information access or not; it is extremely difficult to have it both ways, limited to some and not to others. It is inherent in the theory of how a computer works that if encoded information is made public, partial limitation on access or use collapses.

Virtual electronic mail networks

Over the past decade and a half there has been established an effective global electronic mail system—without carrier or government sanction, without planning, and without even a fully agreed-upon data transfer protocol. This network is defined by its various unpublicized and difficult-to-document international interconnections, mostly non-common carrier, yet linked with all the common-carrier data services. It generally works well even though it has no directory, no map can be drawn of it, it collects no direct revenue, and getting access requires some knowledge of communications networks and the interconnection nodes between computers. Though its services are illegal in some jurisdictions and ignored in others, its official non-existence makes it impossible to monitor.

Electronic mail systems use a variety of equipment, software, interfaces, and operating parameters. Interconnection takes place via a commonality called the American Standard Code for Information Interchange, an international alphabet set, and through a rapidly evolving and changing set of protocols and links. There is no single accepted version of these codes. Machines ranging from microcomputers to mainframes are connected in different ways to the public switched telephone and data networks via dial-up modems, via answering modems, via direct links to packet networks, or direct links through private circuits, or via all of these means. These systems behave as units through the application of flexible, essentially expert-systems software for code translation, routing, and message transfer in conjunction with programmable interface devices.

Transparency of the network

This virtual net is growing rapidly, yet many of its heavier users are not even aware of its existence. While it has no name, it has many components: the U.S. Defense Department's ARPAnet, EDUCOM's BITNET, USENET, CSnet, and more commercial insignia such as MCImail, ITT Dialcom, GTE Telemail, British Telecom Gold, The Source, CompuServe, and still others. And the system's invisibility is increased by the myriad virtual, yet interconnected, networks set up by private firms and special interest user groups.[5]

That the operation of this virtual net is confusing is obvious, and it also confuses any attempt at network control. Messages, sometimes encrypted, are transferred in bulk quantities through an ever-changing array of virtual store-and-forward facilities across national boundaries and across corporate boundaries, despite rules and regulations that were intended, for tariff and other reasons, to regulate such information transfer.

INTELLECTUAL PROPERTY

None of the proposals to modify national or international copyright conventions has been able to come to grips with fundamental paradoxes of the stored-program computer. Just as it will be difficult to confine information resources to arbitrary geographic boundaries on global, distributed computer networks, protection of information that flows across national or corporate borders will become increasingly impractical for any purpose. Some examples of the difficulty in ensuring such protection follow.

Power of the program

The computer is mathematically defined by its program, not necessarily by its electronic architecture, and computer programs are made up of multitudes of other programs. Therefore separating

5. *Information Technology Newsletter* (Office for Technology Information, Harvard University, Cambridge, MA) (Feb. 1983 and Sept.-Oct. 1984).

novel work from work in the public domain, or separation of natural processes from new art, often requires prohibitively complex evidentiary efforts.

Internal and external copying constraints

All digital machines copy in order to work at all. Copying takes place millions of times per second from buffers to arithmetic or logic units, from registers to memory, and from memory stores to communications latches for further copying down the telecommunications line. Switching nodes operate in this same manner and so does every other element in a digital network.

Computers are duplicators *extraordinaire,* and in this environment there is no way for an originator of electronic information to have control over what mode of reading or copying may take place somewhere down the line—ever. The nature of computer processing is such that at any one time there may be multiple copies of text or data residing in various parts of the machine or spread all over a multimachine, multiowner, multijurisdiction distributed network.

Electronic publishing

Especially for those involved in scientific research, electronic mail is a form of publishing more powerful in its potential impact than any other, paper-based form previously available. Information transmitted in this way may be easily stored, modified, and/or transmitted and retransmitted with speed and outreach heretofore unimaginable. But how to assure proper attribution of authorship or document claims of first discovery—ethical principles zealously guarded by the academic community?

Electronic data bases

What should be done about data bases full of raw numerical or other forms of data that are combined with data from disparate, separately owned sources? What if the data are constantly changing, or are being changed recursively by some program under the control of artificial intelligence?

Most of the international legal regime with respect to data bases has assumed a static form of electronic data base technology, almost an analogue of the printed book. With centralized mainframes accessed at slow speed, using so-called dumb time-sharing terminals, that paradigm may have made sense. But with artificial intelligence programs that take data and text from different sources and use them to make inferences or create quite novel outputs, without the source data ever being presented in any human-readable form, this view of data bases will be totally obsolete. Such programs already exist.[6]

Software engineering

Microchips are the creative outputs of software engineering, yet older copyright and patent theories pretend that there is a difference between lines engraved by silicon compilers—or created by electron or laser beams—and machine or object code from conventional computer compilers.[7] These semantics are

6. *Science*, 24 Feb. 1984, p. 802; 23 Mar. 1984, p. 1279; 27 Apr. 1984, p. 372; 15 June 1984, p. 1225; and 10 Aug. 1984, p. 608.

7. Allen Newell, "The Models Are Broken, the Models Are Broken!" *University of Pittsburgh Law Review*, 47(4): 1023-36 (Summer 1986); Solomon, "Intellectual Property"; R[ichard] J[ay] S[olomon], "Evolving International Software Protection: New U.S. Rules Confront the Real World," *International Networks Newsletter*, 19 July 1985;

not germane to the real world of computer science. Pretending things are different when they are not will only lead to distortions in legal regimes. Will a software writer gain greater protection if the software is engraved rather than magnetic? Does engraved data on laser disks that control electrical processes qualify for chip protection? What if the physical process is optical rather than electrical? How can we know the difference?

Piracy

The ultimate test of copy protection is how well a copy can be made by a pirate. With digital data storage the answer is, perfectly! Compact audiodisks, videodisks, digitized software can all be copied without degradation. Attempts at encryption do not work well in practice, for encryption complicates disk access immeasurably. But more important, at some point the data must be decrypted either to work in the computer memory space or to be presented to a human. Whatever that point is, a virtual or physical daemon can be placed to intercept the clear data.

THE SHIFTING LOCUS OF CONTROL

The use of computers in telecommunications networks began as a direct, though more sophisticated, replacement for simple sequential switching. Similarly, the application of telecommunications to computers initially was to bring their computational power closer to users. Today, however, the computer or switch is readily programmable to perform numerous functions, and any networked processor can perform switching as well as diverse processing, from desktop micros to mainframes. These technological facts would be insignificant and trivial in a discussion of policy options, save for the implication they have for the ability to transcend former network barriers and, by extension, for what virtual networks mean to sovereignty.

Where will control reside?

With the application of literally tens of millions of these computer-based systems—perhaps hundreds of millions as each white-collar and blue-collar worker and each robot and robotic component gain virtual network addresses—a global integrated system will have been achieved. This is not to imply that the industrial world will be integrated because of computer communications; far from it. What it does mean is that information will be accessible to those entities that understand how the network works and, indeed, to those that create and manage these virtual networks to their own ends.

The same technology that makes it difficult to control information once it is on a network can, if turned around, also be used to partition data. But it is not the owner of the network hardware who manages and creates these new partitions or virtual boundaries. Instead it is the users—the software writers, to be exact—who now have greater control. "Virtual" is the operative word when discussing computer networking. What we are approaching is a new form of industrial state, with shifting boundaries as resources are created, used, and linked; with indistinct boundaries as industry itself shifts and reacts to changing times and technologies; and with

F. Anceau and E. J. Aas, eds., *VLSI Design of Digital Systems* (New York: North-Holland, 1983), esp. pp. 387-433 and passim.

boundaries quite incongruent with older political borders. As with the heavily scientifically utilized BITNET architecture of today, maps will be hard to draw and will cease to have much meaning.

The future of common-carrier communications in an all-digital, virtual-networked computer environment raises a difficult set of issues—for science, for industry, for government, and for humanity at large.

ANNALS, *AAPSS*, **495**, January 1988

Analysis and Reanalysis of Shared Scientific Data

By THEODOR D. STERLING

ABSTRACT: Enormous strides in data handling and in telecommunications technology have made possible multiple access and sharing of data files. Sharing of research data often is resisted, however, because of (1) potential loss of monetary, political, or psychological reward; (2) potential conflict and disagreement; and (3) potential exposure of extreme bias or fraud. Found acceptable have been (1) the distribution of very large data files for further study, such as Public Use Tapes from the Bureau of the Census or the National Center for Health Statistics; and (2) the use of large data aggregates that serve as very comprehensive catalogs, such as bibliographic files for literature searches. Instances of scientist-to-scientist data sharing depend on individual arrangements. There are, however, many recorded instances of refusal to give access to data. Data that are liable to lead to conflict or controversy seem to be shared only if they fall within the realm of Freedom of Information legislation or court orders.

Theodor D. Sterling is professor in the School of Computing Science, Faculty of Applied Sciences, at Simon Fraser University, Burnaby, British Columbia, Canada. His interdisciplinary training was in quantitative biological and social sciences and statistics at the University of Chicago and Tulane University. He was an early worker in the uses of the computer in the life sciences, has published extensively in that area, and has published on the social impacts of computing on democratic processes and on the humanizing and dehumanizing dimensions of information technology.

WHEN asked by the editor of this volume to produce a concise article on how computer-based network technology will influence collaboration among scientists far removed from each other and each other's physical facilities, a number of colleagues and I accepted the accompanying charge to "examine the positive and negative consequences of the shift away from modes which have heretofore characterized the conduct and communication of science."

At first, I did not fully appreciate that this charge grants to technology the power to affect modes that characterize the conduct and communication of science so that these modes change and somehow become different. It occurred to me later, however, but only after I agreed to do the editor's bidding, that I do not really agree with this thesis of technology's inexorably and inevitably affecting the modes of conduct and communication of science or of other human activities. There is yet a second thesis to consider, one that has ample reason to be explored in the present context: that the way scientists do things modifies the technology that is used to support the things scientists do. With respect to the characteristic conduct of communication in science, this second thesis calls for an examination of the evidence showing the extent to which available technology—in the broadest sense—is being shaped and modified to conform to the modes of conduct that characterize collaboration and communication between scientists.

In short, there are two opposing views: first, that changing technology forces change in the conduct of scientists; and second, that scientists tend to select that technology that continues to support established practices of collaboration and communication. Thesis one is implicit or explicit in almost all investigations into the impact of modern technology, not only on science but on just about everything. Thesis two proposes that there are certain strong economic, psychological, and political forces that guide how scientists communicate with one another and with formal media of scientific communication, and that these forces tend to select, among those available, technologies supporting dominant economic and political interests. The motivating force supporting thesis two aptly is called reinforcement politics.[1]

Of course, neither thesis one nor thesis two needs to rule the roost. The impact that modern technology has on the conduct of scientists and the impact of the forces that shape the conduct of scientists vis-á-vis the selection of technology depend on circumstances. Sometimes the impact of technology may dominate; at other times the impact of reinforcing politics and economics may dominate; and sometimes the model may be mixed, elements of both impacts intruding.

THESIS ONE: THE EASING OF DATA SHARING

The revolution in data handling and, with it, multiple access to data, rests on the same principle as the computer revolution itself. It rests on the ability to store information in machine-readable form and to use that information to direct and control a series of steps so as to guide some procedure from a beginning to an endpoint.

The history of scientific computation, which encompasses data processing, is a history of the development of number

1. James N. Danziger et al., *Computers and Politics* (New York: Columbia University Press, 1982).

systems, algorithms, calculation aids, and, to a lesser extent, means to store symbols and codes. A fairly complete record exists of the search for suitable methods of data processing beginning with the invention of hieroglyphics. These methods were reinforced by the early mechanization of accounting, using counting tables, starting with the Egyptians, and counting machines made of bronze, starting with the Romans.[2] The introduction of positional notation—Arabic numbers—led to the invention of logarithms and appropriate procedures that made large-scale calculation possible, and this was paralleled by the invention of machines to help in the performance of these calculations. The slide rule is not exactly unknown to scientists of my age bracket.

The technology that lagged behind was that of automation in recording, reading, and storing symbols. Only with the development of machines that could read symbols, store and manipulate or transform them were new vistas of data processing opened. With the success achieved in building machines that could read and be set and reset by symbolic codes, all factors determining automatic data-processing methods were now united in the modern computer. These rapid developments—of symbol systems, procedures for processing, hardware and software—now created the ability to manipulate data in machine-readable form from whatever source, store them, and send them from one point to another by telecommunications machines. Even the methods for accessing and manipulating data, for whatever purposes, now are in machine-readable form.

2. Michael R. Williams, *A History of Computing Technology* (Englewood Cliffs, NJ: Prentice-Hall, 1985).

While it always has been possible for consenting scientists to collaborate and share data, actually to do so was extremely laborious before now. Modern technology has eased access to data, both for the primary investigator and for whoever may want to share in this access.

Software support for automation of data sharing

Data are stored in some orderly fashion through the mediation of appropriate software. Insofar as data are stored in machine-readable symbols on a particular medium—which may be tape, disk, or bulk storage, for example—they have to be committed in a form in which individual items can be extracted. Structuring a file to ensure recovery of wanted data components is an important topic in computer science. In order to ensure proper placement of relevant data and security against errors, input to a well-controlled and well-constructed data file is mediated via specially designed input software. At the same time, software controlling input places an immediate and obvious limitation on any collaboration. The collaborator can only add to the data file if the collaborator has access to the input software.

Input-regulating software is related to software for data recovery. Investigators think in terms of parameters such as height, weight, distance, or density rather than in terms of positions of binary digits on a tape. A physicist counts collisions as a function of time or material density and a biologist correlates heights and weights. But the observation of a collision, the density of materials, the height, weight, or other data placed in storage are stored as machine-readable symbols and are assigned to specific locations on whatever storage medium is used. Rather than requiring an investigator to go through the laborious process of specifying which data fields are to be extracted and inserted in a proper spot in some analysis program, the investigator is provided, as part of the data storage system, with software by which a request for a particular data item can be made using a name or label as a tag. This request then is translated into a series of instructions that extract whatever symbols are to be found at locations relevant to the data items that bear a particular name or label. Further, an investigator may not be interested in all data in a category, such as height, but only in a subclass, such as the heights of male infants younger than eight months. A good data-oriented software package enables an investigator to specify not only data fields, but data fields conditional on other data fields. Again, software controlling extraction of data places an immediate and obvious limitation on collaboration. A collaborator must either know the layout of the data and have the facility and skill to write data-extraction programs, or know and have access to the software that directs data extraction.

Both distance and speed of transmission create cost problems. To a large extent, these problems have been solved by networks of communication channels. These channels make it possible for a user in a remote location to be connected with a local exchange and pay only the costs of the actual transmission, which is just a tiny fraction of the length of time spent on a terminal in an interactive process. Transmission costs can be eliminated entirely by sending data tapes or data streams to a collaborator. Investigators then work independently on their data files and exchange modifications or information as needed. The use of telecommunications networks is reserved mostly for instances where the cost of creating and maintaining a large but often used data file is distributed among users.

Types of data-sharing activities

As the previous section suggests, no insurmountable barriers to the sharing of data are created by demands for suitable software. As a result, instances of shared data do exist, and these appear to be of three types:

—instances where a copy of the data is transmitted from one investigator to another. All investigators who have a copy of the original data process these data independently. They may communicate with each other in the course of their work. They basically exploit the same data, but in their own way;

—instances where investigators have access to a central file containing a large body of orderly, cataloged information but without the ability to modify or add directly to the central file; and

—instances where investigators have access to a central data file to which they may contribute as well.

Examples of data files made available *in toto* are Public Use Tapes obtainable from such agencies as the U.S. Census Bureau or the National Center for Health Statistics. These public-use files contain data resulting from enumerations, surveys, and related kinds of activities by the data-collecting agency. Data that are made publicly available are purged of sensitive information, which mostly comprises items that could disclose the identity of informants. Public Use Tapes may also contain intermediate variables created by some of the analysis programs of the data-collecting agency apart from the original data. The files are usually well documented because frequently their contents have already undergone extensive processing.

New material may be added to a Public Use Tape in two ways. The originating agency may announce that it has a modified version of the tape available for distribution or, under certain circumstances, an individual investigator may ask to have additional information introduced for the purposes of a particular analysis. If that information is deemed to be confidential, the investigator may send a tape with part of the data needed for the analysis and have the originating agency provide missing data or perhaps even analyze the sensitive data and provide the requesting investigator with the results.

Possessing a copy of the data file. One great advantage of having a copy of the data file is the intensity with which analysis may be carried out. But extensive analyses of large data files are costly and, despite decreased transmission costs, are still best performed locally. Large data aggregates, such as those compiled by the U.S. Census Bureau or the National Center for Health Statistics, provide the raw material for testing hypotheses of many kinds and, indeed, hundreds of papers are published in scientific journals that are based on analyses utilizing publicly available versions of those data. Investigators who have special problems may obtain additional related data files and combine these in a linked arrangement. Such linked arrangements may very much enlarge the reach of scientific analysis. It may be possible, for instance, to combine hospital records containing diagnosis and date of entry and discharge of patients with information about pollutant levels during the same periods obtainable from the air pollution network.[3] Such analysis between linked data files is not practical, however, unless an investigator controls all the files to be linked.

Access to a central file. The best example of a central data file that scientists may utilize for extracting information without having the ability or need to enter data of their own is the use of data banks, such as MEDLINE, TOXLINE, or CANCERLIT, containing references pertaining to specific scientific literatures. These data banks function as very extensive catalogs. The user enters these data banks, using a particular language in which requests are made. Relatively simple programs then sort and search through files of categorized information. A search through a catalog of published materials yields titles, authors' names, abstracts, keywords, and a list of references corresponding to the demands of the inquiry. The usefulness of these data banks of published materials is attested to not only by the large number of such files in existence, but also by the creation of special systems such as the one known as DIALOG, which can address and search through a large number of bibliographic reference files.

Data-file modification. I do not know of a publicly described example of interactive data sharing among widely separated contributors who both use and enrich a common data aggregate. There are instances, however, of commonly used collaborative data aggregates. One

3. Theodor D. Sterling, Seymour V. Pollack, and James J. Weinkam, "Measuring the Effect of Air Pollution on Urban Morbidity," *Archives of Environmental Health*, 18:485 (Apr. 1969).

example is the Building Performance Database (BPD).[4]

BPD is a large aggregate of data about buildings and their occupants. It contains information extracted from various building studies about the type of building, the location within a building complex that was studied, various aspects of the building's engineering components—especially its ventilation system—measured contaminant levels in the building, occupant complaints, and any other information deemed relevant in studies of building performance. The various studies furnishing information about these buildings were usually done by agencies such as the Occupational Safety and Health Administration (OSHA), Public Works Canada, and the Centers for Disease Control, or by insurance companies or universities. The information in BPD at the present time is stored at the Simon Fraser University Computer Center. Access is possible through any number of datanets. The software package for extracting data is very much expanded to include some of the necessary preliminary data analysis, such as the development of tables or lists. There is a software package for inputting data. There are plans to make the software package available to collaborating institutions; however, at present input capabilities are not passed on to collaborators.

Technical problems in implementing access

Experience with the three types of shared data systems does not indicate the existence of insurmountable obstacles to full data sharing. But obstacles do exist. Shared data systems require an inquirer who is highly qualified professionally and either acquires needed skills or has assistance for communicating with the software. Another obstacle is cost.

Level of user sophistication. The degree of communication skill required usually is grossly underestimated. Languages that allow effective manipulation of complex data bases are necessarily complex. To some extent, an illusion of the ease of extracting data is created by the type of information system that provides a prepackaged display—usually in response to pressing a key on a keyboard. For instance, in preparing an information base on possibly toxic exposures and what to do about them, the Canadian Centre for Occupational Health produced the following type of simple, fast, and user-friendly procedure.

Lists of topics are presented to the inquirer who, by selecting among successive lists of categories and chemicals, converges upon the particular chemical to which there was an exposure. The information pertaining to prevention, treatment, or palliation of contact with that chemical then appears on a screen or is printed out. This type of list-driven information system, however, is more akin to a sophisticated set of catalogs than to the analysis of data archives for the large variety of questions constituting scientific inquiry.

Level of training and experience. To analyze an archive that contains data in some complex structure requires a capability to group, sort, convert, and prepare data. The user must be offered an extensive data dictionary—usually identifying symbols and individual data items

4. Elia M. Sterling et al., "Building Performance Database," *10th CIB Congress* (Washington, DC: International Council for Building Research, Studies and Documentation, in press).

by acronyms—and a set of instructions—that is, a language—with which the data in the archive may be selected and manipulated. Often, analytic programs for making tables, testing for statistical independence, performing log linear analysis, curve fitting, or graphing are part of the instruction package.

Time to learn and practice. A good manipulating language, one that can effectively provide the user with many different services in data manipulation, requires time and effort to learn. The more useful the instruction set, the more services are made available through such a data-base manipulating language, the more complex the grammar of such a language gets to be. Commands for data manipulations are a further complication because they need to contain names of data entities. The data themselves are usually structured in some relational scheme and organized in a dictionary of data names, often containing many hundreds of entries. Thus the manipulator not only must know the language but also must be familiar with the data dictionary, unless names of data entities are to be looked up in the dictionary each time they are needed—a lengthy procedure. Manipulative commands must be carefully entered because there are many data names, and a single typographical error in a string of statements may prevent processing. Therefore, a good data-management language provides for the storing of chains of commands and their modification via an editing subsystem. That useful service also enlarges the set of instructions, however. Thus an adequate instruction set is extensive and requires learning and practice and sometimes considerably greater time commitment on the part of the user than a busy investigator can afford to devote.

Availability of technical assistance. The potential data sharer may have to learn more than an information system's manipulation language. Extensive service programs are embedded within the larger operating system that regulates the activity of the total computer facility. For instance, BPD is embedded in an operating system called the Michigan Terminal System (MTS) at Simon Fraser University. Simple extraction of data does not represent an obstacle. Scientists who not only wish to extract data but also perform analyses on them, however, are forced to learn the idiosyncrasies of MTS as compared to other systems. Also, the fact that the BPD information system has software written for a facility that uses MTS creates severe difficulties for interactive processing with other facilities that do not operate under MTS. There are only a limited number of installations that use MTS, which complicates interactive sharing no end. A scientist may have considerable difficulty in finding skilled help in the university computing facility for interactive work with an operating system not in use at the scientist's institution.

Multiple inputs. One obstacle not yet faced by BPD—nor, to our knowledge, by any other scientific data system—concerns the ability of collaborators to input information to the data base. If scientist A analyzes a file to which scientist B adds data, problems of communication are created. These problems are generally treated as a serious difficulty in commercial applications, such as in an airline reservation system where a number of terminal-using reservation clerks may seek to allocate the same seats, at the same time, to different customers. No good practical solutions have been found for scientific systems that would permit collaborative input, although this lack may be due more to the paucity of systems that experiment with on-line sharing than to the existence of any fundamental technical barriers.

Cost. One cannot discount the cost of creating and sharing data aggregates. Lack of adequate funds places an obvious limitation on the exploitation of available technology. So, for instance, the Canadian Heritage Inventory Network will eventually contain information on the what and where of archaeological sites, historic sites, treasures, paintings, and other artifacts. So far, however, only the inventory of paintings and treasures is complete enough to support scholarly research. In fact, many municipal and provincial art galleries have access to the Canadian Heritage Inventory Network through hard-wired terminals. Similarly, the Southwestern Archeological Research Group failed to establish a data bank because the funds were not forthcoming to support the categorization of massive archaeological data in the American Southwest. Even if it is reasonable to believe that the encoding of all relevant scientific data will proceed despite high cost, the process will be much slower for some disciplines than for others.

THESIS TWO: POLITICS OF SHARING

The sharing of data has been an issue in science ever since the creation of knowledge has been a human enterprise. The high-sounding principle to which all scientists pay obeisance, that science is public and its data are public, is not always honored in practice. The ability of scientist A to link to scientist B's data is influenced by a large number of economic, social, psychological, and political factors that have determined in the past, and will very likely continue to do so in the future, the willingness or unwillingness of one scientist to give access to data to another scientist.

Statistics on how often access to data is demanded or refused are hard to come by. There is no public scorekeeper to keep tally. Knowledge of incidents of refused data sharing is largely anecdotal.

Incidents concerning data on health

Leon Kamin reports refusal of access to data in studies that were at odds with the scientific literature on the genetic basis of schizophrenia.[5] Access to data was first refused on the grounds that divulging the data about subjects would not be in line with the condition under which information from the subjects had been solicited. When Kamin pointed out that subjects' identities could be protected, the investigators refused access because they had not finished analyzing their data. These dame excuses were offered by the American Cancer Society when a committee of 10 senior scientists requested access to the data of the Million-Person Study. Subsequently, this refusal led to an exchange of letters in *Science*, and, indirectly, to a conference called by the National Institutes of Health.[6] The National Cancer Institute was concerned with when to release data and with being forced by the release to share information on findings that re-

5. Leon J. Kamin, *The Science and Politics of IQ* (New York: Halsted Press, 1974); S. J. Ceci, "Scientists Attitude toward Data Sharing" (Paper delivered at the annual meeting of the American Association for the Advancement of Science, Chicago, IL, 17 Feb. 1987).

6. Theodor D. Sterling, "Access to Data," *Science*, 173:676 (1971); Ellis Blade, "The Public Access to Science," *Science*, 175:123 (1972); Theodor D. Sterling, "Scientific Data: Public or Private," *Science*, 177:651 (1972).

lated to possible carcinogenic properties of commonly used substances or substances that were to be introduced into common use.

Attitudes of scientists

In the spring of 1985, Ceci and Walker surveyed 790 researchers from a variety of physical and social sciences.[7] The survey was not meant to be a random selection of scientists, but Ceci and Walker did compare groups of investigators, such as biotechnologists who stood to reap large rewards from certain findings, with social scientists who were not likely to participate in a reward structure to the same degree. Ceci and Walker found few significant differences in opinion between scientists from various fields. The rate of refusal to share data ranged from 16 to 20 percent. There are indications, however, that the percentage of refusal may be higher in actual practice. In a second survey that replicated the results of the first, Ceci and Walker asked scientists to describe their own experiences in trying to obtain data from colleagues or to comment on their associates' attitude regarding data sharing. The majority reported that their colleagues were not prone to sharing data, even data obtained with benefit of federal funds.[8]

Obstacles to data sharing

There appear to be three major reasons for the underlying lack of enthusiasm to share data:

—potential loss of monetary, political, or psychological reward;
—potential for disagreement and/or outright conflict; and
—potential exposure of extreme bias or fraud, whether deliberate or through carelessness.

Loss of reward. Possession of data may have financial and personal rewards. There is a child research institute in San Francisco that inherited contact with a large cohort of children and is following that cohort as the children grow older. The cohort consists of a large sample of children whose parents' various life-style factors were recorded at birth. That institute for years was the sole possessor of these data, which were used to justify long years of support by the National Institutes of Health for follow-up studies. The research group refused all requests for parting with any of their data.

When Anna Freud gave access of Sigmund Freud's letters to Masson, she set up a situation that not only enabled Masson to become a leading spokesperson for a particular school of interpretation of the causes of hysteria but also offered Masson the opportunity to write a best-selling book, serialized in the *New Yorker* before publication.[9]

Ceci points out the ample rewards to biotechnologists for many discoveries. Certainly, engineers have mightily prospered—and more so the companies for which they work—from discoveries, many of which were made with the support of public funds. The American Cancer Society has utilized smoking and health studies to solicit support for its work against smoking. On the other hand, the American Cancer Society has had relatively little to say about forms of pollutants or of toxic contaminants,

7. S. J. Ceci and E. Walker, "Private Archives and Public Needs," *American Psychologist*, 38:414 1983).

8. S. J. Ceci, "Scientists Attitude toward Data Sharing."

9. Jeffrey Masson, *The Assault on Truth* (Toronto: Collins, 1984).

especially in occupational settings, other than those that issue from tobacco.

Disagreement and conflict. A second source of resistance to data sharing is the expectation of disagreement or conflict or both. Data are very often requested in the course of adversary processes involving regulating agency hearings, special commissions, administrative law judges, and civil litigations—and sometimes these adversary processes involve all four.

One example is a recent study by Blair et al. of the health effects of formaldehyde on an industrial population.[10] OSHA has been concerned that the current industrial formaldehyde standard might be too high. Industry strongly denies that there are findings of health effects possibly associated with exposure at presently acceptable levels. For reasons difficult to understand, the Blair et al. study was carried out jointly with the Formaldehyde Institute and a number of major formaldehyde-using industries. Perhaps because of industry participation, the reported results gave a clean bill of health to industry.

The study design, conduct, and conclusions were bitterly attacked during 1986 OSHA hearings. Simultaneously, the Blair et al. conclusions were most relevant to a class action suit brought against producers and the government of Canada by residents of Quebec whose homes had been insulated with urea formaldehyde foam insulation by a federally subsidized program. With support provided by l'Office de protections du consommateur of the Province of Quebec, data for this study were requested from the National Cancer Institute and reanalyzed. It turns out that, where Blair et al. did not find an association with lung cancer, a different categorization of the data and a more thorough control for confounding variables yield the opposite—namely, a significant increased risk for lung cancer for the industrial cohort study.[11]

This finding resulted in the reopening of OSHA hearings.[12] Thus access to the Blair et al. data had far-reaching results. But the incident is not an isolated one. In another example, a review of the U.S. Veterans Study and reanalysis of its data found errors of programming that had led to misclassification of 50,000 out of 250,000 subjects.[13]

A particular problem. A matter of special sensitivity—and one that fuels the fear of conflict—is the possibility that almost always exists of using the same data to obtain different results. Data can be massaged—that is, subsamples of data can be emphasized, other subsamples neglected—in such a way that not one but a number of different conclusions can be based on the same data set. The more data are collected, the greater the opportunity for massaging. Thus access to a particular assembly of data from a large-

10. Aaron Blair et al., "Mortality among Industrial Workers Exposed to Formaldehyde," *Journal of the National Cancer Institute*, 76(6):1071 (June 1986).

11. Theodor D. Sterling and James J. Weinkam, "Reanalysis of a National Cancer Institute Study on 'Mortality among Industrial Workers Exposed to Formaldehyde,'" Report to l'Office de protection du consommateur of the Province of Quebec (submitted for publication, 1987).

12. Federal Register, 29 CFR Part 1910, Occupational Exposure to Formaldehyde 51, No. 239 (12 Dec. 1986), 44796.

13. Theodor Sterling and James J. Weinkam, "What Happens When Major Errors Are Discovered Long after an Important Report Has Been Published?" (Talk delivered at the annual meeting of the American Statistical Association, Washington, DC, 16 Aug. 1979).

scale study is sometimes of considerable interest to adversaries where conflicts of a doctrinal, political, economic, or psychological nature are already afoot.

Bias or fraud. A third motive for withholding data is the existence of bias, especially of bias bordering on fraud, if not fraud itself. Many scientists take an activist role concerning various issues, and some may use their position and acknowledged expertise to present findings that are relevant to a contentious public issue. But sometimes conclusions publicized by some scientists who also play a role as advocates rest on stubborn belief rather than acceptable analysis of data. There may be a wide difference between conclusions actually supported by the data and those published by the author. While some scientists may justify their biased presentations by their fervent belief in the justice of their cause, they still find it prudent to refuse access to their real findings. Then, of course, there are instances of outright scientific fraud, and these have become of growing concern to the scientific community in recent years.

A FINAL WORD

Data can be stored in a machine-readable form, and modern telecommunications systems make it possible for other scientists to gain access to these data. There are technical problems, some of which have been and some of which need to be resolved so as to give full access to the stored information, but none of these technical obstacles appears to be insurmountable. Providing that all the technical prerequisites are met, modern telecommunications technology, combined with advances in computing, facilitate the sharing of data so that sharing becomes practical and inexpensive. Data sharing also is seriously advocated as a desirable extension of scientific conduct.[14] At the same time, the potential facilitation of data sharing impinges on areas of scientific activity that have been marked more by conflict than by cooperation.

One reasonable conclusion from our brief review is that the politics and economics of the interaction of scientists would tend to encourage some and discourage other uses of the technology for different aspects of data sharing. A number of aspects of data sharing have proved to be useful and acceptable, and they will continue to flourish. Foremost among them are two: the use of large data files for reanalysis, such as files made available by the Census Bureau or data sets of archaeological significance, or the Public Use Tapes of the National Center for Health Statistics; and the use of data aggregates that serve as very comprehensive catalogs, such as for literature search, or collections of technical data, as in BPD.

But the idea of scientist-to-scientist cooperation more likely depends on individual arrangements, and on the whole, the prospects for that particular cooperation to flourish are dim. Economic motives, motives of personal power, possible disagreement, prospects of conflict, likely detection of bias or fraud—and who among us is not conscious of some aspects of bias in and fraudulent exaggeration of the importance of our work?—all combine to discourage data sharing.

This does not mean that some data will not always be available for review by others. Legislation, such as the Free-

14. B. J. Yolles, J. C. Connors, and S. Grufferman, "Obtaining Access to Data from Government-Sponsored Medical Research," *New England Journal of Medicine*, 315(26):1669 (Dec. 1986).

dom of Information Act, or orders issued by a judge might compel some scientists working for a public service organization or engaging in expert testimony under some litigation to allow their data to be scrutinized by others. It is worth noting that, in the United States, the social and economic science branch of the National Science Foundation has recently adopted a policy of requiring all recipients of research grants to archive any data set produced under a grant in a credentialed data library such as the archival facilities of the Inter-University Consortium for Political and Social Research at the University of Michigan or the National Technical Information Service, a branch of the Department of Commerce. In another recent development, the Committee on National Statistics of the prestigious National Academy of Sciences issued a report that reviews the benefits and costs of sharing data and urges individual investigators to share data as a regular practice, this among a host of recommendations directed at journals, agencies, and academic institutions as well as the individual researcher.[15]

These are encouraging signs, perhaps even signals of change on the horizon. It is to be kept in mind, however, that the National Science Foundation possesses no powers of enforcement and that the promulgation of codes of ethics by committees—even prestigious ones—is no guarantee of conformance by the research community at large. It will be some time before the effectiveness of these recent developments can be assessed. Meanwhile, it is our opinion that the politics of data sharing are such that they will discourage rather than reinforce that desirable practice.

15. S. E. Fienberg, M. E. Martin, and M. L. Straf, eds., *Sharing Research Data* (Washington, DC: National Academy Press, 1985).

ANNALS, *AAPSS*, **495**, January 1988

Public Policy Concerning the Exchange and Distribution of Scientific Information

By FRED W. WEINGARTEN and D. LINDA GARCIA

ABSTRACT: The new information and communication technologies hold tremendous promise for both carrying out scientific research and disseminating the results. The importance of these technologies to the achievement of economic and military ends may, however, hamper their use for scientific purposes. Similarly, and perhaps of greater significance, the way these technologies are employed in defense and in the private sector may adversely affect the progress of science and seriously constrain the whole realm of scientific activity. Thus conflict over the use of new technologies is likely to give rise to a number of information-related public policy issues, issues on which the scientific community will have to take a stand. Decisions made in this regard will govern the flows of information within and across the borders of the United States, profoundly affecting scientific communication and perhaps even determining the future of U.S. science.

Fred W. Weingarten is program manager of the Communication and Information Technologies Program, Office of Technology Assessment, U.S. Congress. He is responsible for policy studies done for congressional committees in the areas of telecommunications and information systems.

D. Linda Garcia is a senior analyst at the Office of Technology Assessment, U.S. Congress. She has worked in the areas of information technology and education, information policy, information technology research and development, and intellectual property rights.

TODAY, a wide variety of information and communications technologies are rapidly being developed that are profoundly altering the nation's and the world's communication systems. Included among them are storage technologies such as optical disks, computer networks for distributed information processing, fiber optic transmission systems, digital telephone switches, teletext and videotext systems and other forms of electronic publishing, and satellite distribution systems. These technologies not only permit the introduction and development of a wide variety of new information-based products and services; they also provide much greater flexibility in the processing, packaging, and distribution of information.

These rapid technological advances are bringing with them major social and economic changes. They are changing the way people work and conduct their business, how they interact and relate to one another; the way they learn, create, and process information; and their needs and expectations. In fact, these new technologies are altering the way that people view themselves and their places in the world.[1]

Together, the development and use of these new technologies are helping to usher in what some social observers characterize as a postindustrial or information society.[2] In this society, the creation, use, and communication of information plays a central role. Not only will the amount of information continue to increase, but people will also rely on it more and in different circumstances. The changes brought on by the new technologies will often generate new social, economic, and cultural opportunities and choices, which will bring with them the need for major policy decisions.

Given the capabilities of the new information and communications technologies, and the enhanced role and value of information in all areas of life, disagreements have arisen over how, by whom, and for what purposes these technologies should be used. The creation of new opportunities for some often generates problems for others. For example, using new technologies to process and store socioeconomic data about individuals, public agencies can improve their efficiency and effectiveness; in doing so, however, they may create problems in the protection of individual rights to privacy.[3] Similarly, individuals can use the new technologies to access information more easily and inexpensively, but, in the process, they may undermine the existing system for protecting and enforcing intellectual property rights.[4]

1. Sherri Turkel, *The Second Self: Computers and the Human Spirit* (New York: Simon & Schuster, 1984).

2. For discussions and characterizations of the information society, see, for example, U.S., Congress, Office of Technology Assessment, *Computer-Based National Information Systems*, OTA-CIT-146 (Washington, DC: Government Printing Office, 1981); Susan Artandi, "Man, Information and Society: New Patterns of Interaction," *Journal for the American Society for Information Science* (Jan. 1979); Daniel Bell, *The Coming of Post-Industrial Society* (New York: Basic Books, 1973); James R. Beniger, "Information Society and Global Science," this issue of *The Annals* of the American Academy of Political and Social Science.

3. U.S., Congress, Office of Technology Assessment, *Federal Government Information Technology: Electronic Record Systems and Individual Privacy*, OTA-CIT-296 (Washington, DC: Government Printing Office, 1986).

4. U.S., Congress, Office of Technology Assessment, *Intellectual Property Rights in an Age of Electronics and Information*, OTA-CIT-302 (Washington, DC: Government Printing Office, 1986).

One use of the new technologies that might cause considerable conflict is their use in science. The new technologies hold tremendous promise for both carrying out research and disseminating the results. As the following discussion points out, however, as we move further into an information age, with increasing competition for knowledge and information, the use of information and communication technologies for scientific purposes may conflict with and hamper their use for economic or military ends. Similarly, and perhaps more important, the way these technologies are used in the private sector and in defense may significantly alter the nature of scientific activity. To understand how such conflicts may occur, it is necessary to look first at what is meant by science.

THE NATURE OF SCIENTIFIC ACTIVITY AND ITS RELATIONSHIP TO SOCIETY

Science can be defined as a social activity whereby people use rational thought to achieve empirical ends.[5] Viewed as such, it is clear that scientific activity is common to all societies, even to preliterate ones. For, as Bernard Barber has pointed out, "the germ of science in human society lies in man's aboriginal and unceasing attempt to understand and control the world in which he lives by the use of rational thought."[6]

Although present in all societies and cultures, science has evolved and progressed—more slowly to begin with and then by leaps and bounds—throughout the course of human history. A dynamic activity that develops in response to continued, critical analysis, science has become increasingly conceptual and generalizable over time.[7] As part of its evolution, it has developed special investigative procedures and specialized roles.[8]

Science and society interact so that the nature and development of science at any particular time is strongly affected by the societal context in which it is operating.[9] The values and practices of some societies have been more conducive to scientific progress than others. The heavy emphasis on magic in preliterate cultures, for example, was incompatible with rational, empirical knowledge, and thus it inhibited its development.[10] Similarly, the Oriental conviction that the nature of the universe was unknowable long discouraged the practice of science.[11] Then, in contrast, during the Middle Ages, the writings of Saint Thomas Aquinas gave science a tremendous boost. The religious belief in the rationality of God, and hence in the rationality of nature, made the idea of science a real possibility.[12]

Science also benefited greatly from European developments during the sixteenth and seventeenth centuries: the birth of Cartesian philosophy, the invention of differential calculus, appreciation for the importance of theory, the development of new tools and procedures for observation and experimentation, mercantile capitalism, and the Protestant ethic.[13] More recently, the values of the liberal state—rationality, progress, free-

5. Bernard Barber, *Science and the Social Order* (New York: Free Press, 1952), pp. 7-8.

6. Ibid., p. 7.

7. Ibid., pp. 45-46.

8. Talcott Parsons, *The Social System* (New York: Free Press, 1951), p. 333.

9. Ibid., pp. 332-33.

10. Ibid., p. 333.

11. Barber, *Science and the Social Order*, p. 46.

12. Ibid.

13. Ibid., pp. 50-52.

dom, and individualism—have strongly favored the growth of science.[14]

Although a reflection of society, science is also an activity unto itself, with its own moral purpose, structure, procedures, and norms.[15] Devoted above all to the advancement of knowledge, science typically prescribes the acceptable means and procedures for attaining it.[16] The scientific community, moreover, monitors itself, requiring its members to adhere strictly to four essential norms. These are (1) universalism, or the requirement that claims of truth be judged according to cognitive, and not personal, criteria; (2) disinterestedness, the requirement that scientists put the goals of science above the desire for personal gain; (3) organized skepticism, requiring scientists to evaluate all claims of truth in terms of empirical and logical criteria; and (4) communism, requiring that the findings of science be freely communicated and shared with the public.[17]

The publication and dissemination of scientific results are critical activities in the scientific process, if science is to function in accordance with these norms.[18] Publication and dissemination are necessary not only to meet the requirement of communism; they are also the means by which the criteria of disinterestedness and organized skepticism are fulfilled. Scientists are able to sublimate their need for pecuniary rewards because they gain fulfillment through the recognition that they receive from their colleagues for having contributed to the advancement of knowledge. Were their works to remain unpublished, they would have no reward.[19] So important is this kind of reward to the scientist that "once he has made his contribution, [he] no longer has exclusive rights of access to it. It becomes part of the public domain of science."[20]

Benjamin Franklin exemplified this ethic. Explaining why he turned down an offer from the governor of Pennsylvania to patent the Franklin stove, he wrote to a friend:

> I declined from a Principle which has weighed with me on such occasions, vis. That as we enjoy great Advantages from the Invention of others, we should be glad of an opportunity to serve others by an invention of ours, and this we should do freely and generously.[21]

Nor did scientists traditionally seek to market their discoveries. Louis Pasteur's attitude was typical. Although he himself estimated that the use of his method would save Fr100 million per

14. Ibid., pp. 85-87.

15. Ibid., p. 84.

16. Robert K. Merton, *The Sociology of Science: Theoretical and Empirical Investigations* (Chicago: University of Chicago Press, 1973), p. 270.

17. Ibid., pp. 270-78.

18. The practice of publishing and disseminating scientific works was one of the developments that provided an impetus for science during the Renaissance. As Barber points out, "The [scientific] societies also became the channels not only of national but of international communication in the new knowledge. Each society had regular foreign correspondents charged with reporting events in his country; and reading letters of these correspondents was a feature of the meetings." Barber, *Science and the Social Order*, p. 54.

19. Merton, *Sociology of Science*, p. 293. Describing the process, Merton says, "This way in which the norms of science help to produce this result seems clear enough. On every side the scientist is reminded that it is his role to advance knowledge and his happiest fulfillment of that role, to advance knowledge greatly. This is only to say, of course, that in the institution of science originality is at a premium." Ibid.

20. Ibid., p. 294.

21. As quoted in Bruce Willis Bugbee, *Genesis of American Patent and Copyright Law* (Washington, DC: Public Affairs Press, 1976), p. 72.

year, he was not interested in profiting financially from his discoveries. As he explained to Napoleon III, "In France scientists would consider they lowered themselves by doing so."[22]

Without the publication and dissemination of scientific work, moreover, the scientific community would have no objective means by which to maintain the quality and validity of research. As Robert Merton has pointed out:

> Science is public and not private knowledge; and although the idea of "other persons" is not employed explicitly in science, it is always tacitly involved. In order to prove a generalization, which for the individual scientist, on the basis of his own private experience, may have attained the status of a valid law which requires no further confirmation, the investigator is compelled to set up critical experiments which will satisfy the other scientists engaged in the same cooperative activity. This pressure for so working out a problem that the solution will satisfy not only the scientist's own criteria of validity and adequacy, but also the criteria of the group with whom he is actually or symbolically in control, constitutes a powerful social impetus for cogent, rigorous investigation. The work of the scientist is at every point influenced by the intrinsic requirements of the phenomena with which he is dealing and perhaps just as directly by his reactions to the inferred critical attitudes or actual criticism of other scientists and by an adjustment of his behavior in accordance with these attitudes.[23]

The four scientific norms identified by Merton, which were long supported by society at large, have been quite successful in promoting the rapid advancement of science in the United States. Today, however—and in part, as a result of these scientific developments—society is undergoing a number of changes that may very well undermine these norms. In particular, given the greatly enhanced value of information in all aspects of life, pressure is mounting from many quarters, and for a variety of different purposes, to restrict the flow of scientific information by limiting access to it. Nowhere is this process better illustrated than in the following two cases, one involving the increased commoditization of knowledge and information and the other, the restriction of the dissemination of scientific information for purposes of national security.

INCREASED COMMODITIZATION OF KNOWLEDGE AND INFORMATION

The new communication and information technologies will play a greatly enhanced role in all aspects of life. In fact, their availability and use may, in many cases, be the critical factor in personal and organizational success. The enhanced value of these technologies is reflected, first of all, in the growing number of people who, from whatever realm of life, are striving to integrate these technologies into their daily activities and operations. It is reflected, moreover, in the greatly increased market for information-based products and services and in the flourishing of new industries to provide for these burgeoning information needs.

Not all of these technological opportunities, however, will be exploited. In fact, taking advantage of some opportunities may preclude the development of

22. As quoted in J. D. Bernal, *Science and Industry in the Nineteenth Century* (Boston: Routledge & Kegan Paul, 1953), p. 86.

23. Robert Merton, *Science, Technology, and Society in Seventeenth Century England* (New York: Howard Fertig, 1938), p. 219, as quoted in Harriet Zuckerman, "Deviant Behavior and Social Control in Science," in *Deviance and Social Change*, ed. Edward Sagarin (Newbury Park, CA: Sage, 1977), pp. 88-89.

others. The potential for conflict is likely to be most pronounced in areas, such as science, where the economic value of information is very high. For it is under such circumstances that the discrepancy between the need for exclusions and the need for distribution, sharing, and use is the most starkly drawn.

In contrast to the world of science, the value of information from the perspective of the economic realm is in its exclusivity—that is to say, in the ability of its owners to be able to exploit the difference between what they know and what other people do not know. In a horse race, for example, the value of an accurate assessment of the horse's chances increases directly with the exclusivity of that wisdom, and the value is obviously decreased by sharing. Similarly, an important factor in encouraging investment is the presumption that the investor is better informed than others about the outcome of the enterprise. To the degree that all investors have equal access to information, this potential for difference is reduced, along with the incentive for investment.[24]

The tension between the norms of business and the norms of science is clearly evident today in many institutions of higher education. Once the undisputed center of research efforts in the United States, American universities are today competing strenuously with one another and with business and governmental research institutes for money and resources. In the present economic climate, most universities are finding it difficult to compete. Equipment for advanced scientific research is extremely expensive to buy and to maintain. Faculty members, drawn by the superior research opportunities and financial benefits offered by private firms and government, are leaving the universities and taking their research teams with them.[25]

To make universities more competitive and more financially independent, many universities and colleges have established new relationships with the business community. Such arrangements generally involve some form of joint research. Probably the first such arrangement is the 10-year Harvard-Monsanto agreement signed in 1975, which provided that the chemical firm Monsanto would provide Harvard Medical School $23 million to support the research of two professors.[26] Many others have followed. Increasingly popular with both business and academia, these agreements are considered to be mutually beneficial. The universities receive money to finance research and to replace obsolete equipment, while the business communities obtain—often on a proprietary basis—access to basic, advanced technological research.[27]

24. Christopher Burns and Patricia Martin, *The Economics of Information*, Contractor report prepared for U.S., Congress, Office of Technology Assessment (Boston: Christopher Burns Inc., 1985).

25. W. R. Lynn and F. A. Long, "University-Industrial Collaboration in Research," *Technology and Society*, 4:199 (1982). The rapid obsolescence of laboratory and research tools, combined with the highly complex and sophisticated nature of the equipment now needed for advanced technology research, results in capital costs beyond the reach of most academic institutions. For example, in 1970 the cost of new instrumentation in U.S. universities' laboratories was estimated to be $200 million; in 1980, it was $1 billion. National Research Council, *Revitalizing Laboratory Instrumentation* (Washington, DC: National Academy Press, 1982).

26. David Dickson, *The New Politics of Science* (New York: Pantheon, 1984), p. 66.

27. Philip L. Bereano, "Making Knowledge a Commodity: Increased Corporate Influence on Universities," *IEEE Technology and Society Magazine*, pp. 8-9 (Dec. 1986).

Industry representatives are also actively courting the traditional scientist-scholar to leave academia for jobs in industry. As one professor of biological science at Harvard University explained, "At this point it is mind boggling. I'm courted every day. Yesterday, some guy offered me literally millions of dollars to go direct a research outfit on the west coast. . . . He said any price."[28] Such offers have caused much soul-searching among research scientists. While some respond favorably to these developments—even to the point of creating their own firms to exploit their discoveries for profit—others have opposed them as unsuitable for academic science. Trying to sort out what is appropriate behavior for academics and academia, a number of major universities have themselves begun working together to develop policy guidelines for university-industry relationships.[29]

Critics of new alliances between industry and academia are concerned lest these new partnerships serve to undermine the traditional norms of academic and scientific research. Above all, they fear that such agreements will be proprietary in nature and thus inhibit the exchange of research. Regretfully noting such developments, Donald Kennedy, president of Stanford University, has said, for example:

> The commercial environment is characterized by many more constraints upon the openness and accessibility of scientific and technical information than is the university environment. Proprietary restraints on the free exchange of data have already begun to crop up. . . . There are at least three or four incidents during this past year (1980-81) at scientific meetings, at which the traditional evaluation of research had always been expected to prevail; there were communications in which a scientist actually refused on questioning to divulge some detail of technique, claiming that, in fact, it was a proprietary matter and that he was not free to communicate it.[30]

This problem is exacerbated by the fact that, as the market value of information increases, so does the pressure to treat information and knowledge as economic commodities. Not surprisingly, rivalry for ownership is becoming increasingly common at institutions of science and research where the potential for profits is very high. The claims and counterclaims of ownership are continually multiplying: claims of students against students, students against faculty members, faculty against faculty, and the university against students and faculty.[31] A particularly contentious issue is work for hire. Some university administrators now argue, for example, that just as companies automatically own the copyright on works done on company time and with company resources, so too universities should have the rights to everything created in conjunction with their facilities.[32] At Virginia Polytechnic Institute, this policy has been carried so far that lawyers at the university have

28. As quoted in Henry Etzkowitz, "Entrepreneurial Science and Entrepreneurial Universities in American Academic Science," *Minerva*, 21(2-3):199 (Summer-autumn 1985).

29. "Academe and Industry Debate Partnership," *Science*, 219(4481):150-51 (Jan. 1983).

30. L. Lindsay, "Troubled Conscience in Academe: Industry's Help Priced Too High," *Christian Science Monitor*, 29 Oct. 1982, as quoted in Bereano, "Making Knowledge a Commodity," p. 12.

31. Dorothy Nelkin, *Science as Intellectual Property: Who Controls Scientific Research* (New York: Macmillan, 1984), pp. 1-8.

32. Ivars Peterson, "Bits of Ownership: Growing Computer Software Sales Are Forcing Universities to Rethink Their Copyright and Patent Policies," *Science News*, 21 Sept. 1985, pp. 189-90.

recently concluded that students' assignments are the property of their professors.[33]

Conflicts such as these are likely to become more intense, more complicated, and more difficult to resolve as we move further into an information age. Only recently, for example, Harvard biologist Walter Gilbert announced his company's intention to copyright the sequence of human deoxyribonucleic acid.[34] Although most of the leading molecular biologists in the United States have some connections with industry, many of them view Gilbert's statement of intent with considerable alarm. They fear that if proprietary rights are granted to the mapping and sequencing of the human genome, the search for medically important genes will be greatly retarded. In addition, they themselves are asking whether there might not be certain kinds of information that, given their overwhelming importance to humankind, should never be owned by anyone or distributed through the marketplace.[35]

Notwithstanding such concerns, proprietary restrictions on the distribution and flow of information are likely to become commonplace in an economic environment that is increasingly fueled by information production and use. The repercussions of such developments will be felt not only by those who cannot afford the price of knowledge and information but also by the community of science as a whole. As Thorsten Veblen once said, "The outcome of any research can only be to make two questions grow where one question grew before."[36] Without the exchange of research, there can be no questioning, and without questioning there is no dynamic, no driving force, to science.

NATIONAL SECURITY AND SCIENTIFIC INFORMATION

For centuries, science and technology have been recognized for their contribution to the ability of societies to make war. Among his many and varied creative activities, Leonardo DaVinci invented military hardware. The invention of the stirrup transformed warfare in medieval Europe by establishing the technological basis for a horse-mounted armored fighter. More recently, in the United States, the National Academy of Sciences was formed during the Civil War to help the Union war effort, and the National Research Council was similarly a product of World War I.

Yet it was during World War II, with inventions such as radar and, of course, the atomic bomb, that the contribution of basic science to military power was fully recognized. For the bomb did not originate from weapons laboratories but from formulas on the chalkboards of theoretical physicists. It was then also, with the realization that science could make an immediate and profound contribution to war, that the modern conflict between national security concerns and the open exchange of scientific information was fully joined.

Legislative control over information flows

The Atomic Energy Act of 1946 defined a category of information called "Restricted Data" as information subject

33. Ibid.

34. "Who Owns the Human Genome?" *Science*, 237(4813):358-61 (July 1987).

35. Ibid.

36. Thorsten Veblen, "The Evolution of the Scientific Point of View," in *The Place of Science in Modern Civilization and Other Essays* (New York: Viking, 1919), as cited in Barber, *Science and the Social Order*, p. 21.

to government control. The act was written in such a way as to allow the inclusion of information not created by or in the direct control of the government. In certain cases, according to the definition, Restricted Data could be designated to be "Born Classified"—it could be confiscated and classified by government no matter who developed or held it. Thus was scientific publication in certain fields placed under threat of government restriction, even if it concerned basic research results and even if the research was conducted privately.

Subsequently, numerous other laws, executive orders, and regulations have established controls over the dissemination of scientific information. The Arms Export Control Act and its International Traffic in Arms Regulations, the Patent Secrecy Act, and the Arms Export Administration Act, as well as executive directives, have been used at times by the government to attempt to block the transfer of scientific communication. In the last two decades, government officials have invoked these and other rules to try, with varying success, to stop the publication of articles, the presentation of papers at conferences, the participation by foreign graduate students in academic seminars, and access to government-supported research computing facilities.

The first such efforts to become widely known and followed in the press were attempts in the mid-1970s to control the publication and patenting of results from research projects funded by the National Science Foundation involving cryptography.[37] Since then, the controversy has ebbed and flowed, as compromises have been negotiated to avert criminal prosecution of offending scientists or court tests of constitutionality.[38] The most recent attempt at compromise centers on the concept of fundamental science as defined in National Security Decision Directive 189. In it, fundamental science, on which controls will not normally be imposed, is "basic and applied research in science and engineering, the results of which ordinarily are published and shared broadly within the scientific community, as distinguished from proprietary research."

The circularity of the definition, however, illustrates the difficulties in dealing with this issue. That is, fundamental research—research not subject to control—is defined as research not usually subject to control. Furthermore, given our earlier point that the line between appropriable and nonappropriable research in the commercial sector is blurring, the definition does not rest on a very robust distinction. In fact, the definition ties the two areas of conflict together in an interesting way. It suggests, plausibly, that once the research community gives up open communication of research on economic grounds, it loses force in its argument against national security controls.

Pressures created by military needs

Despite recent compromises between researchers and government, government interest in controlling the flow of scientific information on national security grounds is bound to continue, for the conflict is rooted deeply in the evolution of the information society discussed in the beginning of this article.

37. Fred W. Weingarten, "Controlling Cryptographic Publication," *Computers & Security*, 2(1):41-48 (1983).

38. David A. Wilson, "Federal Control of Information in Academic Science," *Jurimetrics*, 27(3):283-96 (Spring 1987).

In the first place, as the development of the atomic bomb suggests, the distinction between basic science and engineering and technological products of military interest is blurring. This convergence is true today not only in nuclear physics but in such diverse fields as biophysics, optics, geology, oceanography, mathematics, and computer science.

Second, the military dependency on so-called dual-use technologies—technologies with both civil and military applications—is also expanding. Armies have always needed boots and trucks, and navies have always needed navigational information similar or identical to that used by commercial sailors; the overlapping interest of military and civil sectors in certain technologies is not new. The overlap, however, seems to be expanding as the military becomes dependent on the same technological infrastructure that underlies civilian life. There seems to be little obvious dual-use technology underlying a bomb or rifle, but today the military uses many of the same microelectronics technologies that are used in consumer products, as well as aeronautical and nautical design and construction technologies that have civilian applications.

Finally, the U.S. Defense Department is a major supporter of research in the nation's laboratories. In some fields, the Defense Department funds the majority of research pursued by civilian scientists, including research that is not explicitly for military purposes.[39] This direct involvement by the military serves to strengthen the presumption in some minds that there is a significant security interest in the results of the research. Furthermore, because the work is government funded, controls over the publication of results, if imposed, are seen more as the appropriate by-product of a contractual arrangement than as an abridgment of freedom of speech, because the sponsor has some proprietary rights over the disposition of the results. Yet when the Defense Department funds a significant majority of the research in a field, as it does in some cases, such rights, if exercised, can chill publication in an entire body of research.

It would be unfair to suggest that this debate is purely between the scientific community and defense interests. There have been times when the scientific community has worried deeply about the implications of research it was pursuing and wondered if publication, or even the research itself, should be suppressed. On the other side, a defense establishment that depends heavily on science and technology cannot afford to take actions that would stifle U.S. research. The conflict is much deeper—intrinsic to the information society. How can we continue to pursue science in the open, public international forum it requires while protecting secrets that are vital to our security?

THE PROMISE AND THE DILEMMA

The challenges faced by science due to the pressures our modern information society places on open communication are more far-reaching than the conflict between those asserting a right to control and those asserting academic freedom and First Amendment rights to publish. Certainly, this is a central issue. But another, possibly even more important, question is whether science, under such

39. With respect to computer science support, see U.S., Congress, Office of Technology Assessment, *Information Technology R&D: Critical Trends and Issues*, OTA-CIT-268 (Washington, DC: Government Printing Office, 1985).

pressure, can avail itself of new opportunities that modern communication and computer technologies offer.

Just as the printing press shaped during the Renaissance what we know as modern scholarship, the electronic information tools emerging will most likely have, we think, a major transformational influence on scientific research. Such views are not idiosyncratic. The impact of information technology on research has been the subject of congressional hearings, study projects of the National Academy of Sciences, and reports to the National Science Foundation. The National Science Foundation's current program developments in supercomputers and networking reflect a growing awareness that information technology is becoming a fundamental instrumental infrastructure for the conduct of scientific research in many ways.[40] Among the predictions in this literature are the following:

1. New generations of supercomputers, particularly the more affordable minisupercomputers, will provide large amounts of computational power for simulation and data analysis.
2. Optical disks and other technologies will provide inexpensive and transportable media for storing vast amounts of data.
3. Worldwide high-speed data communication networks will link researchers with each other and with specialized computers, data bases, and software.
4. High resolution, real-time color graphics will allow scientists to visualize the outcome of their theories and/or experiments in new, more intuitively suggestive ways.
5. Software based on artificial-intelligence techniques will help researchers screen enormous amounts of information for new patterns and ideas and bridge gaps between diverging specialties.

Taken together, these predictions suggest that scientific research in the next century will be as dependent on electronic information technology as it was in the past centuries on libraries and microscopes.

The technology also appears to facilitate scientific communication greatly, making control even more difficult. Controlling access to a supercomputer attached to an international communication network is no easy task. Compact optical disks small enough to fit in a shirt pocket can carry billions of bits of data. Data communication networks can transmit large software packages and data sets anywhere in the world in seconds. Hence information technology appears to support strongly the openness and international character of scientific research.

To control the flow of scientific information, then, may require controlling access to and even use of these modern research instruments. Such policies may seem feasible at first glance, for governments still have a great influence over communication policy. Furthermore, the cost of some of the technology will be great enough that only government will be able to underwrite it, and with the funds could come strings. At last word, the National Science Foundation was still in negotiation with its university-based supercomputer centers over access

40. For a particularly provocative but informed picture of the future prospects, see John Seeley Brown, "The Impact of the Information Age on the Conduct and Communication of Science," *Information Hotline*, pp. 18-23 (June 1986). See also U.S., Congress, House, Committee on Science and Technology, *The Impact of Information Technology on Science: Science Policy Study Background Report No. 5*, 1986.

to the facilities by foreign students.

If access to modern information technology becomes synonymous with the ability to engage in scientific research, control of access to that technology, in the name of national security, economic advantage, or other social objectives, will determine who can participate, both within the U.S. research community and internationally. Furthermore, because the technology seems inherently to favor openness, attempts to control scientific information flow may require policies so draconian as to force science to forgo some of its most exciting opportunities in the next century.

The scientific community cannot hold itself aloof from the issues. It must play in the game. The decisions made in our information society will govern the flows of information within and across the borders of the United States and will affect scientific communication profoundly—perhaps even determining the future of U.S. science.

ANNALS, *AAPSS*, **495**, January 1988

Science Sublanguages and the Prospects for a Global Language of Science

By ZELLIG HARRIS and PAUL MATTICK, Jr.

ABSTRACT: Scientists have limited access to results published in languages in which they are not fluent. One solution to the problem is suggested by some results of investigation into the nature of language generally and the language of various sciences in particular. The information provided in language is given not only by the meanings of individual words but also by the relations among words, especially by the regularities of their co-occurrence. Particular sciences, furthermore, are characterized by particular sets of such relations among words. These relational structures are shared by discourses within the same scientific field in different languages; these structures can thus be seen as expressing the information carried by language in the field irrespective of national language. Because the informational structures are discoverable in a computable way, the solution suggested here to the problem of international communication in science would at the same time provide facilities for the computer processing and retrieval of scientific information on a large—potentially a global—scale.

Zellig Harris is Benjamin Franklin and University Professor of Linguistics at the University of Pennsylvania. He is the author of numerous books, including Methods in Structural Linguistics, Mathematical Structures of Language, *and* A Grammar of English on Mathematical Principles.

Paul Mattick, Jr., teaches philosophy at Bennington College and is a fellow of the Institute for the Humanities at New York University. He is the author of Social Knowledge.

THE development of science in the modern period coincided with the growth in importance of national linguistic boundaries in cultural life. As signaled by Descartes's presentation of his *Meditations* and certain of his scientific works in French as well as in Latin, this reflected, inter alia, a new secular order of ideas and institutions in opposition to the traditional order represented by Latin as the language at once of classical philosophy and Christian theology. The new cultural order took form in the context of the formation of modern nation-states in Europe, in which societies and academies of the sciences and arts displaced the earlier international system of universities, which were themselves soon to be transformed.

The general issue of European political-cultural disunity was very much the inspiration for the first great project of an international language of science, Leibniz's plan for a *Characteristica Universalis*, a symbolic representation of conceptual elements calculational operations on which would resolve all disputed questions. Leibniz himself wrote in French and German as well as Latin, which remained a basic language of science until the nineteenth century, when the ever more rapid pace of scientific development within national university systems, often in close connection with industrial development, led to its abandonment. On the other hand, certain areas of research became identified with particular languages, so that, for example, students of organic chemistry were obliged to learn German in order to read important research results.

The idea that international understanding would be fostered by a universal language lay behind a number of attempts at inventing such a language, of which Esperanto has been the most significant. Interlingua was invented in 1951 for use at scientific and medical meetings, but it has had little impact, partly as a result of being based on English and Romance languages only. At the present time, English is the closest to an international language of science, due largely to the economic and political dominance of the United States. But the bulk of scientific work is published in many national languages. This limits the access of scientists to results published in languages in which they are not fluent. At the same time science remains by nature an international and transcultural enterprise. The continuing explosion of scientific research around the world makes the question of a global language of science an important one. Considering this explosion of scientific research and the facility that advanced communications technology imparts to scientific interchange, the possibility of a global language of science becomes a reasonable one to examine.

LANGUAGE AND INFORMATION

In each area of science, and more generally in many specific subject matters, the use of language is limited in particular ways—and limited in the same ways no matter what language is being used. This is why it is easier to translate scientific texts than literary ones. These limitations of use, and the interlanguage similarity of the limitations, are due to an essential property of language.

This property is that the information provided in language is given not only by the dictionary meanings of individual words but also by the relations among words, especially by the regularities of their co-occurrence, or combination in sentences. When the grammar of a language is described in its most compact and essential form, it is found that every contribution to the structure of a sentence—which words combine in what grammatical relation—makes a fixed contribution to the meaning of the sentence. This is an underlying form-content relation not altered by grammatical transformations, which change the form of a sentence but not its information—for example, the reduction of "I prefer for me to leave last" to "I prefer to leave last" does not change the information imparted.

LANGUAGE AND STRUCTURE

An important specialization of the form-content property is the sublanguage structure. It has been found that the use of a language in the texts or talk of a reasonably well-structured subject matter, especially a science, is limited in ways that go far beyond the limitations of ordinary grammar. In ordinary language, sentences consist of verbs with nouns—or whole sentences—as subject and in many cases also as object, with very few hard and fast restrictions as to which nouns can be subjects or objects of which verbs. Thus "child" may be a much more frequent subject of "sleep"—as in "The child slept"—than is "chair" or "universe"; but the latter cannot be excluded from the grammar—as in "That chair slept for years in the attic" and "Until the Big Bang, the universe slept."

In scientific writing, in contrast, we find sharp restrictions on word co-occurrence. In biochemistry, for example, one can say, "The polypeptides were washed in hydrochloric acid," but "Hydrochloric acid was washed in polypeptides," while a grammatical English sentence, cannot appear in a biochemistry article. For each science we find particular sets of nouns that can occur as subject, or object, of particular sets of verbs, to make not just a general noun-verb-noun sentence type as in English or French but a family of distinct sentence types, each with its particular subsets of verbs and of nouns.

LANGUAGE AND SUBLANGUAGE

In any system of a mathematical type, if there is a subset of the system that is closed under operations of the system then that subset is called a subsystem of the whole. "Closed" here means that an operation on any member, or pair of members, of the subset yields another member of the subset.

The subset of English sentences found in texts of a science has this character; grammatical operations on a sentence of the science will produce another sentence that can occur in texts of that science. For example, the active form of "The polypeptides were washed in hydrochloric acid" is "We washed the polypeptides in hydrochloric acid," which is also a sentence of biochemistry. Similarly, the active of "Hydrochloric acid was washed in polypeptides," which is not a biochemistry sentence, would be "We washed hydrochloric acid in polypeptides," which would also not be found in a biochemistry article. The set of English sentences in biochemistry, or in some subfield thereof, constitutes a sublanguage of English.

SUBLANGUAGE FORMULAS

A further linguistic property makes those previously mentioned relevant to the problem of international scientific communication. In every language in which there are texts and conversations in biochemistry, there is a biochemistry sublanguage, and so for every such field.

If we examine the structure of, for instance, the biochemistry sublanguage of French and the biochemistry sublanguage of English—that is, the subsets of nouns, verbs, and other elements and the various sentence types made of them—we find that they are in all essentials identical. If we mark the various word subsets in the English biochemistry sublanguage by letters—for example, by using *P* for "polypeptides" and other molecules that might be treated by washing, *W* for certain laboratory operations, and *S* for certain solutions—we could represent the sentence types by sequences of these word-class symbols. Such a sequence would be "PWS" here. We can show that the same symbol classes and sentential symbol-sequences suffice to characterize the word classes and sentence types of the French biochemistry sublanguage. This means that articles in whatever language in the given biochemical field could be represented by sequences of the same types of formulas. Starting with a science sublanguage, expressed in the words of one language or another, we have reached a science language expressed in symbols.

WHAT THE FORMULAS REPRESENT

The importance of the formulas is not that they are reminiscent of mathematics or chemistry. Indeed, a universe of interformula relations defined a priori, which is at the heart of mathematical equations and chemical-reaction formulas—and such as Leibniz dreamed of for his *Characteristica Universalis*—does not exist for science languages. The science-language formulas are more like the formulation of numbers in a particular notation such as the customary decimal expansion, or like the formulas for each individual chemical compound.

The science-language formulas have two major properties, however. One is that, like any fixed representation, they locate each item under discussion in preset positions relative to other items. If we want to know about any particular object or interobject relation studied in a field, we know where to look for it in the formulaic representation of any document or sentence. The other property of the formulas is that they allow us to free the representation of information from the noninformational features of language. Many languages have grammatical requirements that can go beyond what is needed to express information. For example, English requires that each verb carry a tense—say, present, past, or future—even in cases where tense is irrelevant to the information carried, as in the case of general statements or universal laws such as "Two plus two equals four." The formulas dispense with everything except what is relevant to the information that is distinguishable in the given field. It is therefore not surprising that the same formulas represent the same information irrespective of the language used.[1]

A NATURAL SCIENCE EXAMPLE: IMMUNOLOGY

To show what a science sublanguage is like, we present a very brief sketch of the language of immunology research papers circa 1935-66. This was a period when this field was far smaller and more

1. For a detailed examination of a science sublanguage—that of immunology—and a study of the essential identity of sublanguage symbols and symbol sequences in English and French, see Zellig Harris et al., *The Form of Information in Science: Analysis of an Information Sublanguage*, Boston Studies in the Philosophy of Science 104 (Dordrecht: Reidel, forthcoming).

inspectable than it is now and when it had a central research problem of determining which cell was the producer of antibodies. There was a controversy as to whether it was the lymphocyte or the plasma cell, both of the lymphatic system. After it was shown, by electron microscopy and other methods, that both cell types produced antibodies, the controversy was resolved by the understanding that the two cell names pertained to different stages of development of the same cell line. The purpose of the analysis that will be summarized here was to see if one could give a formal representation, in an orderly and usable way, of all the information contained in articles written in this area, if one could locate in the sentence structures—and characterize structurally—the changes in information over the years, and if one could locate and characterize the disagreements between the scientific workers involved.

Analysis of the literature

The study began with a detailed grammatical analysis of each sentence in each of 14 research papers studied, utilizing linguistic methods to recast sentences, where necessary, into forms facilitating the search for patterns of word repetition. For example, passive constructions were transformed into active ones. Words, or groups of words, were considered discourse equivalent when they appeared in the identical linguistic environment; thus nouns found in the context "_______ was injected" were classed together as antigen words. Because the word classes are defined by their occurrence in particular syntactic relations to other words, which thereby fall into other word classes, the procedure yielding these classes simultaneously yields the sequences of them that constitute the sublanguage sentence types.

The immunology sublanguage

In its barest outline, the sublanguage discovered by this process of analysis contained some 15 word classes. The chief ones, each followed here by the capital letter used to represent it in sublanguage formulas, are those for "antigen" (G), "antibody" (A), "inject" (J), "tissue" (T), "cell" (C), "body part" (B); then for verbs occurring between A and C (V; for example, "appear in," "produced by," "secreted by"), verbs occurring between two cell names (Y; for example, "is similar to," "develops into"), and verbs appearing with T or C words (W; for example, "T inflames," "C proliferates"). Words of these classes appeared, combined, in fewer than 10 major sentence types, chiefly those exemplified by "Antigen is injected into body"; "Antigen moves to tissue"; "Cells or tissues change or have some property"; "Antibody appears in cell"; "Cell is the same as or develops into another cell."

Formulaic representation of sentences

The many sets of synonymous words, especially verbs, are considered to be just variant forms of a single word, and the variants are not indicated. Writing each class with its letter symbol, we can represent the information in each sentence by a formula constructed of letters; thus "Antibody appears in lymphocytes" is AVC. The nonsynonymous words within a class are marked by subscripts, as in V_i for "appears in" and, synonymously, "present in," "contained in"; V_p

for "produced by"; and V_s for "secreted by." There are modifiers on certain verbs—such as "not," "increase," the pair "from" and "to," "begin to," and "have a role in"—and on certain nouns—such as "much," "immature," and "family of [cells]." These are marked by superscripts on the word-class letter.

We thus have these major sentence types, illustrated here by generic sentences rather than actual examples from particular papers:

—GJB, for "Antigen is injected into a body part or an animal."
—$GU^{ft}TT$, for "Antigen moves from tissue to some tissue"; the *ft* superscript indicates "from" and "to."
—TW and CW, for "A tissue [or cell] has some property or undergoes a change."
—AVC, for "Antibody appears in, is produced by, or is secreted from a cell."
—CYC, for "Some cell is similar to or is called some cell."
—CY_cC, for "A cell develops into another cell."

In donor research, in which antigen is injected into one animal, and then lymphocytes are injected, or transferred, from that animal to another, with antibodies then being sought in the second animal, an additional sentence type is found: $CI^{ft}BB$, for "Cells are injected from an animal into another animal."

There is a special conjunction, internal to a particular sentence-type sequence, that appears or is implicit in almost all occurrences of the pair GJB and AVC. This is "thereafter" and its synonyms, marked in our formulaic representation by a colon (:). It often carries a time modifier. An example is $GJB{:}^{t}AVC$, for "Antigen is injected into a body part; some time later antibody appears in cells." In inverse order the sentence would read, "Antibody appeared some time after ingestion of antigen." This conjunction takes different grammatical forms, for example, "to" in "The cell contained antibody to the antigen." All these forms synonymously connect AVC—or CW or TW—to GJB.

To recapitulate the analytic procedure

This sketch of the immunology sublanguage is sufficient to indicate the advances in the analysis of science information obtainable from the codification of the sublanguage structure. To begin with, metascience material, which states scientists' relations to the information of the science, can be distinguished from the latter, which appears in the form of nominalized sentences embedded in a recognizable set of contexts, such as "Researchers have shown that _____," "_____ as was expected," or "It was found that _____," as in "It was found that antibody is in lymphocytes," or the equivalent, "Antibody was found in lymphocytes."

We obtain a gross framework for representing the information in the field: the word-class sequence formulas, such as AVC. We also obtain a representation for the specific information in each sentence: the individual formulas with subscripts for different class members and superscripts for modifiers, as in $AV_p^rC_y$, for "Lymphocytes have a role in the production of antibody." The superscript *r* indicates participating in production as against actually producing.

We find, in this particular sublanguage, tightly knit sentence sequences marked by a colon, as in GJB:AVC, for "Antigen injection is followed by antibody appearing in cells." Insertions are possible, as in GJB:GUT:AVC, for "Antigen injection is followed by antigen moving to a particular tissue after which antibody appears in cells." Alternative paths are also possible, as in GJB:TW, for "Antigen injection is followed by a particular tissue being altered." We see how related research lines differ. In the donor research mentioned earlier we have $GJB_1 \because CI^{ft}B_1B_2{:}AVCB_2$, for "Antigen is injected into animal one; thereafter lymphocytes are injected from animal one to animal two; thereafter antibody appears in lymphocytes in animal two." Subscripts here distinguish the two animals.

Some analytical results

Within most papers we find differences in sentence types between the Procedures, Results, and Discussion sections, allowing discrimination between different kinds of science information. Across papers, we can locate change over time. First AVT is replaced by AVC, as attention shifts from whole tissues to cells. Later, a new sentence type, CYC, enters, when more cell types are distinguished and their similarities noted, and when the proliferation of cell names is controlled by saying that some different names are for the same cell. In this connection, we can locate unclarity, as when the proliferation of cell names is not supported by different properties—in the W class—reported for the differently named cells, with the unclarity being finally recognized by CYC sentences stating that these are names for the same cell.

We can locate the disagreements between papers and see their structural status. The disagreements appear as symbol differences at specific points in the formulas. The chief case here is that one set of papers has AV_pC_y, for "Antibody is produced by lymphocytes" or "Lymphocytes produce antibody," while another set has AV_pC_z, for "Antibody is produced by plasma cells," and $AV_p^rC_z$, for "Lymphocytes have only a role in antibody production," and even $AV_p^{\sim}C_y$, for "Lymphocytes do not produce antibody," but does not have $AV_p\,C_y$. The contradiction between $AV_p^{\sim}C_y$ and AV_pC_y is overt.

We can also locate the resolution of this disagreement, when $C_yY_cC_z$, for "Lymphocytes develop into plasma cells," appears in the final papers. Sentences of the form CY_cC, stating that one cell is a later stage of a previous cell, were becoming frequent in the later papers as many cell names and cell-stage names appeared in the course of various experiments. But the two contenders for antibody production, C_y and C_z, had never appeared in the context ——— Y_c———; that is, the development was not recognized as reaching from one antibody-producing cell to the other. When both cells were shown to be producing antibody, the explanation—that they were in the same cell line—was expressed by extending Y_c to the pair of C_y and C_z: $C_yY_cC_z$.

A SOCIAL SCIENCE EXAMPLE: SURVEY INSTRUMENTS

The social sciences are in general not immediately amenable to sublanguage analysis, largely because they are wide-ranging in their topics, and their discussions readily extend into related fields or into examples from daily life. The language of some types of social science survey instruments, however, is restricted in the necessary ways. Within each instrument and within different instruments in one area—for example, that of income and wealth—only a relatively small number of words are used, and they are used in very few combinations with other sets of words. With the word classes represented by symbols, the questions constructed from them can be mapped to symbol sequences, sublanguage formulas.

Application of the method

The analytical procedures applied to immunology texts have in fact been applied to small samples of the instrumentation used in three major national survey series: the Survey of Income and Program Participation, the Panel Study of Income Dynamics, and the National Longitudinal Surveys of Labor Market Experience. Patterns of word-co-occurrence were studied, in order to discover the classes of words, and sequences of these, appearing regularly in questionnaires employed in these series. The main word classes found are for the subject of the question, generally the respondent in the survey; for verbs indicating relation to employment or other income sources; for words for work or other income sources; and for the other categories of information sought by such surveys: duration of employment, amount of pay, and so forth. Three main question types, constructed from these word classes, were found; they appear in the survey questions in various combinations, joined by linguistic connectives to form more complex questions. As in the immunology material, instances of the sentence types may carry modifiers, also of specified types, that qualify the information.

Some analytical examples

The three main question types ask about employment, about welfare program participation and nonemployment income, and about conditions relevant to qualification for program participation. Table 1 shows two examples, representing the information requested by typical survey questions. We give the question formula, a description of each of the constituent word classes, and the question words, put into a standard order for intersentence comparison. The first example, "What was the main reason R could not take a job during those weeks?" typifies questions asking about employment. The second example, "Have you ever received Social Security disability benefits?" exemplifies questions asking about welfare programs and nonemployment income.

In these examples, the word classes are represented by words and mnemonics rather than by single letters, but the principle is exactly the same as in the immunology case. The method of analysis is based only on the occurrences of the words, and not on conceptions of their meanings or any other considerations contributed by the analyst. Nonetheless, this method makes it possible to code, store, and compare the information in sentences and whole documents. For example, once the questions in a group of instruments have been mapped to their formulaic representations, one can easily locate all questions that utilize a particular type of information and also those that utilize these types in particular combinations.

COMPUTABILITY OF THE METHOD

To generalize, an important property of sublanguage structures and science-language formulas is that they are discoverable by the application of fixed procedures of finding the regularities of word combination in a field and not on the basis of subjective judgments or of

TABLE 1
SUBLANGUAGE ANALYSIS OF SOCIAL SCIENCE SURVEY QUESTIONS

Question 1	What was the main reason R could not take a job during those weeks?					
Question formula	SUB	VJOB	JOB	DUR	WHY	
Word-class description	respondent	verb for job	job	duration of job	reason	
Question words in standard order	R	could not take	job	during those weeks	for what main reason	
Question 2	Have you ever received Social Security disability benefits?					
Question formula	SUB	VREC	INC	TYPE	SOURCE	DUR
Word-class description	respondent	verb for recipience	income	type of income	source of income	duration
Question words in standard order	you	have received	benefits	disability	Social Security	ever

semantic properties that lie beyond the capacity of computers. Because of this, computer programs can be developed to represent the sentences of documents in the field by the appropriate formulas and to subject the information represented in the formulas to various sorts of processing. Writing such programs is a daunting task and means adding a major capability to computers. But it has been done, in early stages, for some fields of science and medicine,[2] and also for the social science survey instruments just described.

SUBLANGUAGE COMMUNICATION

We can now consider what this new informational representation means for science communication and for international science cooperation. First, it means that methodologically unified "grammars" of science are possible. That is, there can be languagelike systems in which anything that cannot be said in the science, as irrelevant, meaningless, or grossly nonsensical, is ruled out as "ungrammatical." Second, it means that a computable representation of the specific information in scientific documents—including, in principle, conversations—is a possibility. Third, it means that in each field, scientists of whatever national language are even today speaking a global language, although it is expressed in the sounds and grammatical requirements of their particular languages. That is, while each science language can be viewed as a sublanguage within the spoken national language of each scientist, these languages as used in science communication can also be viewed as just particular pronunciations of the global science language.

It may be possible to overcome some of the communicational differences between the global science-language of a field and its divisive national-language pronunciations. It is unrealistic to expect scientists of whatever national linguistic backgrounds to begin to think, speak, write, hear, or read the statements of their science in formulas. Even in mathematics this is not quite what is done. But there are many ways in which the formulaic representation can be a communicational aid. It would be no great task for scientists to become acquainted with the formulas of their field, once these have been obtained by analysis from texts in that field. Scientists' articles and their papers for international conferences could be accompanied by abstracts or subtitles written in formulas or by formulas for the lead sentences of the main paragraphs. The international nature of the science formula language might also serve to limit the dominance exercised within the world of science by speakers of the leading national languages.

Implications for telescience

Aside from facilitating information processing for the needs of individual scientists and facilitating international cooperation among scientific workers, the fact that the science language is computable offers enhanced capabilities for science institutions. It makes possible computer processing of language data from articles or research reports, including language material added to standard data forms. It also opens a way for the construction and maintenance of data bases and other accessible and processible archives of information, even in real time, beyond what has hitherto been thought possible for computers. It offers certain safeguards on the privacy

2. See Naomi Sager et al., *Medical Language Processing: Computer Management of Narrative Data* (New York: Addison-Wesley, 1987).

and confidentiality of data in processing, as when indications of sources in documents are unseen in computer processing of information in the documents and not in human processing. The fact that differences between languages of origin are irrelevant to the sameness of science formulas means that remote-access multinational archives and data bases can be maintained in real time with little more difficulty than single-language ones, once the translation to the formulas has been worked out for each participating language. In such ways the solution promised by sublanguage analysis to the problem of international communication in science would at the same time provide facilities for the computer processing and retrieval of science information on a large—potentially a global—scale.

This solution would, of course, not suffice to overcome the basic contradiction between the rational and universal character of science, with its implication of the need of all interested human groups for free and equal access to scientific information, and the actual control of science as a political and economic resource by the nationally and socially distinct possessors of social power. The problem that Leibniz experienced is one not of difficulties in communication but of differences in interests. Any development in the direction of freer communication, however, at least points in the direction of a more egalitarian mode of creating and utilizing human knowledge.

ANNALS, *AAPSS*, **495**, January 1988

The Role of Unesco in International Scientific Communication

By JACQUES TOCATLIAN

ABSTRACT: By its constitution, the United Nations Educational, Scientific, and Cultural Organization (Unesco) is concerned with information matters. Under its programs known as UNISIST and the General Information Program, it assists the flow of scientific and technical information across national boundaries, facilitates access to published information and data, and enhances member states' capacity to store, exchange, and use information needed for their development. The main thrust of Unesco's action is along five lines: development of internationally agreed-upon methods, standards, and tools to facilitate systems interconnection and exchange of information; the application of new technologies and the creation of data bases; the establishment of regional cooperative schemes, information programs, and networks; support for the creation of national information policies and infrastructures; and the development of specialized labor.

Jacques Tocatlian studied industrial chemistry in Alexandria, Egypt, textile technology in Milan, Italy, organic chemistry at Utah State University, and library and information science at Drexel University, Philadelphia. He has been with Unesco since 1969, associated with the activities described in this article. He is presently the director of the Division of the General Information Program.

NOTE: The ideas and opinions expressed in this article are those of the author and do not necessarily represent the views of Unesco.

THE United Nations Educational, Scientific, and Cultural Organization (Unesco) has been concerned with information matters since its founding in 1946. Indeed, Article I of its constitution stipulates that the organization shall

> maintain, increase and diffuse knowledge . . . by encouraging co-operation among the nations in all branches of intellectual activity . . . the exchange of publications . . . and other materials of information; and by initiating methods of international co-operation calculated to give the people of all countries access to the printed and published materials produced by any of them.

Over the years, particular attention has been given to issues related to the flow of scientific and technical information across national boundaries, with considerable concentration of concern occurring in the mid-1960s, when the international scientific community became aware of the shortcomings of the existing information services for mission-oriented, interdisciplinary, or problem-solving research and development. It was at the Pugwash Conference on Science and World Affairs in Karlovy Vary in 1964 that it was noted that indexing and abstracting services were developing independently, so that information stored by one of them was not freely exchangeable with information stored by others.

In 1967 Unesco joined forces with the International Council of Scientific Unions to carry out a feasibility study on the establishment of a world science information system, UNISIST, under the supervision of a central committee, chaired by Dr. Harrison Brown of the United States' National Science Foundation, assisted by an advisory panel and various working groups. The details appeared in a feasibility report prepared by Jean-Claude Gardin.[1]

UNISIST: A PHILOSOPHY, A MOVEMENT, AND A PROGRAM

Gardin's report and a synoptic version of it became the working documents of the UNISIST Intergovernmental Conference held in Paris in October 1971 to advise Unesco on the implementation of the feasibility study. It will be noted that the word "UNISIST" was never meant to be properly an acronym, but rather to connote phonetically the part that the U.N. agencies, particularly Unesco, should play in the promotion of an international system for information covering science and technology. "UNISIST" stood for the study, for the con ference, and for the program launched thereafter.

According to the study, UNISIST was to be planned as a continuing, flexible program to coordinate existing trends toward cooperation and to act as a catalyst for the necessary developments in scientific information. The ultimate goal was the establishment of a flexible and loosely connected network of information systems and services based on voluntary cooperation. UNISIST was to be concerned initially with the sciences, applied sciences, engineering, and technology, but it was later to be extended to other fields of learning. It was not conceived as a rigid, predesigned superstructure that would provide the world scientific community with all avail-

1. United Nations Educational, Scientific, and Cultural Organization (Unesco) and the International Council of Scientific Unions, *UNISIST—Study Report on the Feasibility of a World Science Information System* (Paris: Unesco, 1971).

able information in science and technology. Rather, a more pragmatic approach was taken by considering the feasibility from a technical, as well as from a political and economic, point of view.[2]

The broad principles on which the world science information system was to be based are enumerated in the preface of the UNISIST study. UNISIST stands for the unimpeded exchange of published scientific information and data among scientists of the world; hospitality to the diversity of disciplines and fields of science and technology; promotion of compatibility, cooperative agreements, and interchange of published information among the systems; cooperative development and maintenance of technical standards to facilitate the interchange; development of trained workers and information resources in all countries; increased participation of the present and coming generation of scientists in the development and use of information systems; reduction of administrative and legal barriers to the flow of scientific information in the world; and assistance to countries that seek access to present and future information services in the sciences.

Information problems and needs of the 1960s

The information problems felt by the international community were numerous. First, there was the sheer volume of technical reports, articles, and publications, which was increasing at an accelerating rate. Faulty distribution practices and understocked and understaffed libraries made access to this literature difficult; linguistic barriers interposed comprehension difficulties. Less obvious, but more radical, were the changing needs of the scientific community. The interdisciplinary approach to problems required interdisciplinary information. The emerging needs of applied science, technology, and engineering added further complexities. The traditional information services had difficulties in responding to these requirements. New forms of information services seemed indispensable. The development of electronic processing and retrieval systems without provision for their compatibility was leading toward a sort of new Tower of Babel. The lack of adequate infrastructures and trained labor compounded the problems of developing countries.

"If the trends which have created this problem situation are continued without corrective action," said Harrison Brown in transmitting the feasibility study report to the director-general of Unesco and the president of the International Council of Scientific Unions,

> it is the Central Committee's considered judgement that not only national science programs but also science itself will be the losers. With the rising costs of information processing, scientific information will become a luxury which only a few wealthy countries can afford instead of the daily fare of the working scientist. Unless the channels of international communication are kept open and readied to accommodate the ever-increasing volume of traffic, scientists in different countries will work in increasing isolation, unwittingly repeating and duplicating each other's work. Without programmes to focus governmental attention on the needs of countries to develop their information resources, scientists in many countries will work under varying degrees of handicap. This will be particularly true of the developing countries, where the gulf separating their

2. Adam Wysocki and Jacques Tocatlian, "A World Science Information System—Necessary and Feasible," *Unesco Bulletin for Libraries*, 25(2):604-05 (Mar.-Apr. 1971).

resources of knowledge and know-how from those of the developed countries will inexorably widen.[3]

The foundations of an international action

Despite early use of the terminology "world science information system," "UNISIST" emerged from the 1971 Intergovernmental Conference as the name of the long-term program subsequently established in 1972 by the General Conference of Unesco at its seventeenth session.[4]

The General Conference elected a steering committee composed of 18 member states to guide and supervise the planning and implementation of the UNISIST program and authorized the director-general, within the framework of this program:

(a) to undertake activities for improvement of the tools of systems interconnection;

(b) to provide assistance for strengthening the functions and improving the performance of the institutional components of the information transfer chain;

(c) to help in the development of the specialized manpower essential for the planning and operation of information networks, especially in the developing countries;

(d) to encourage the development of scientific information policies and national networks;

(e) to assist Member States, especially the developing countries, in the creation and development of their infrastructure in the field of scientific and technical information.

3. Harrison Brown, "Transmittal Memorandum," in *UNISIST—Study Report on the Feasibility of a World Science Information System*, pp. i-iii.

4. Unesco, 17 C/Resolution 2.131, in *Unesco General Conference Seventeenth Session, Paris, 1972—Resolutions and Recommendations* (Paris: Unesco, 1973), pp. 39-40.

Evolution of the program

The goals of the program have remained constant since its inception. Two areas—development of information infrastructures, and education and training—have maintained their preeminence; the training of information users was added to the latter in 1977. Increasing importance has been given to the promotion of policy formulation and planning at the national, regional, and international levels. One objective, originally called "improving tools of systems interconnection" and later designated as the "promotion and dissemination of information methods, norms and standards," has remained the center of gravity of the program. In fact, UNISIST has since then often been referred to also as the set of internationally developed methods, norms, standards, principles, and techniques governing the processing and transfer of information.

The information needs of the international community for which the UNISIST program was set up have evolved with time. Since its creation UNISIST has increasingly been concerned with scientific information and technological information as these serve the development process. It is this association with development that has provided a new mission orientation for UNISIST as time goes by.

A second conference, the Intergovernmental Conference on Scientific and Technological Information for Development—UNISIST II[5]—was held in Paris in 1979 to review developments since the UNISIST conference of 1971 and to

5. Unesco, *Intergovernmental Conference on Scientific and Technological Information for Development—UNISIST II Final Report* (Paris: Unesco, 1979).

make recommendations for the future. The 1979 conference felt that

> the original recommendations of the 1971 Conference had been sound and were still relevant to the largely changed economic and social conditions. . . . Much had been achieved . . . but there was still a great deal more to be achieved because many countries, especially the less developed ones, still had to develop coherent national policies, set up and coordinate the necessary information infrastructures, and establish systematic programmes for education and training of professional information workers and of final users of information.

The recommendations of the conference were vital in shaping the content of future programs of Unesco in this area.

An important factor in the evolution of the UNISIST program was the longstanding question as to whether Unesco's information programs should be structured by sector or by function. In 1976, the General Information Program (PGI) was created by merging UNISIST with a program concerned with the development of documentation, libraries, and archives.[6] An intergovernmental council composed of 30 member states replaced the former UNISIST steering committee and guided the planning and implementation of PGI. By 1979, at the time of the UNISIST II conference, it was felt that the creation of PGI had brought a number of benefits: it had reduced the number of inconsistencies in Unesco's dealings with member states on matters relating to information transfer; it had brought together experience in infrastructure development and education and training; and it had provided for an integrated approach to information systems planning and development covering libraries and archives as well as scientific and technological information.

REVIEW OF PROGRAM ACHIEVEMENTS

Evaluation of the program results is done at various levels. At the biennial Unesco General Conference, member states' delegations make detailed statements on past achievements, express their needs, and comment on future program proposals. The executive board also provides substantive suggestions and expresses a judgment on the evolution of specific Unesco programs. The Intergovernmental Council for the General Information Program meets every two years to review in detail the results of this program.

A wealth of practical guidance

In 1982, Stephen Parker prepared a guide[7] to selective documents and publications issued by Unesco in this field of activity, giving 104 bibliographic references. Many of the documents and publications referred to are available in more than one language and all are free of charge. In concluding his article, Parker says, "It will be apparent from the foregoing that the guidelines, studies, manuals and other documents issued by Unesco provide a wealth of practical guidance on almost every aspect of information transfer." These documents are announced in the quarterly *UNISIST Newsletter*, which is issued in English, French, Spanish, Russian, and Arabic. In response to requests received from all over the world, some 20,000 documents

6. J. Stephen Parker, *Unesco and Library Development Planning* (London: Library Association, 1985).

7. J. Stephen Parker, "Unesco Documents and Publications in the Field of Information: A Summary Guide," *IFLA Journal*, 10(3):251-72 (Aug. 1984).

a year are dispatched gratis.

Practical guidance is provided, not only through case studies, guidelines, manuals, and other documents, but also by promoting, sponsoring, and supporting such activities as meetings, consultations, and conferences; education and training courses and seminars; study fellowships; pilot projects; consultancy missions; and equipment grants.

It will not be possible to present here the results of years of cooperation with some 160 member states and several nongovernmental organizations in assisting the flow of scientific and technical information across national boundaries, facilitating general access to specialized published information and data, and expanding member states' capacity to store, exchange, and use the information needed for their development. It will suffice to highlight a few examples to illustrate the type of activities undertaken under each of the five subprograms that make up PGI.

Tools for the processing and transfer of information

Scientific and technical information flows from generators to consumers of such information through an information-transfer chain. No single diagram can illustrate the plurality of information services that are in existence. They include publishers, abstracting and indexing services, libraries, clearinghouses, specialized documentation centers, data banks and information networks. They serve different communities, through different channels, following different patterns. Suffice it at this stage to observe that, at each step of the information-transfer chain, the application of internationally agreed-upon methods and standards facilitates the flow and interchange of information. Long and complex negotiations and systematic work have led to the production and use of a number of useful tools. Among them is the *UNISIST Guide to Standards for Information Handling*.[8] This includes an extensive bibliography of 64 pages and analyzes national and international standards, rules, codes, and other normative instruments in the field.

The *ISO Standards Handbook, I: Information Transfer*[9] contains the texts of major international standards in the fields of documentation reproduction, terminology, and library science and documentation.

The *Reference Manual for Machine-Readable Bibliographic Descriptions*[10] serves as a standardized communication format for the exchange of machine-readable bibliographic information between data bases or other information services, including libraries. An adaptation of it exists for the exchange of information on research in progress.[11]

The Common Communication Format, CCF[12] provides a detailed and structured method for recording data elements in a computer-readable bibliographic record for exchange purposes between two or more computer-based systems. A number of national organizations as well as regional ones, such as the

8. Unesco, *UNISIST Guide to Standards for Information Handling* (Paris: Unesco, 1980).

9. International Organization for Standardization, *ISO Standards Handbook, I: Information Transfer*, 2nd ed. (Geneva: International Organization for Standardization, 1982).

10. Unesco, *Reference Manual for Machine-Readable Bibliographic Descriptions*, 2nd ed. (Paris: Unesco, 1981).

11. Unesco, *Reference Manual for Machine-Readable Descriptions of Research Projects and Institutions* (Paris: Unesco, 1982).

12. Unesco, *The Common Communication Format, CCF* (Paris: Unesco, 1984).

Commission of the European Communities, and international ones, such as the United Nations, have adopted the Common Communication Format.

Unesco has organized various seminars on the application of standards and has prepared teaching packages on their use. In addition, it has been concerned with terminological problems, methodologies for compiling scientific and technical thesauri, the presentation of primary documents, and the preparation of scientific papers. It has also established an international system, the International Serials Data System,[13] for the identification of serials. With the recent accession of Australia, India, and China to the International Serials Data System, this network is now approaching 90 percent coverage of world serials production.

Development of data bases

Computerized data bases have been set up in industrialized countries to give users almost instant access to a large proportion of current scientific and technical knowledge by means of data processing and telecommunications technology. It was estimated in the early 1980s that out of the 900 bibliographical and numerical data bases then available on-line internationally, less than 1 percent were produced in the developing countries. The information produced by these countries is vitally necessary for a proper understanding of the various problems affecting people's lives and for developing strategies to overcome them. It is therefore essential that developing countries should be able to use other countries' data bases and set up data bases of their own.

As a result, Unesco has encouraged and supported the application of modern information technology in the creation of data bases. Financial support is provided to representatives from the developing countries to attend relevant seminars, workshops, and conferences. Useful inventories and studies, such as the *International Inventory of Software Packages in the Information Field*[14] and *The Application of Minicomputers and Microcomputers to Information Handling*,[15] have been published and widely disseminated. Three microcomputer-based software packages have been provided free of charge to nonprofit organizations in the developing world. One of these packages is the mini/micro version of the Computerized Documentation System/Integrated Set of Information Systems (CDS/ISIS) software and is developed by Unesco itself. As of March 1987, some 600 requests for these software packages had been filled and there were close to 900 pending requests for CDS/ISIS alone. Training for the utilization of CDS/ISIS is done at the rate of about 100 persons per year. In the framework of PGI, there are about fifty projects for the creation of data bases for which the assistance given comprises the provision not only of software, but also of consultants, equipment, and training. These projects have been carried out in the African countries of Kenya, Mali, Rwanda, Senegal, and Tanzania; in the Arab countries of Tunisia and Syria; in Asia in China, India, Indonesia, Pakistan, Papua New Guinea, the Philippines, Sri Lanka, and Thailand; and in

13. International Centre for the Registration of Serials Publications, *Guidelines for ISDS* (Paris: Unesco, 1973).

14. Unesco, *International Inventory of Software Packages in the Information Field* (Paris: Unesco, 1983).

15. Unesco, *Application of Minicomputers and Microcomputers to Information Handling* (Paris: Unesco, 1981).

Latin America in Brazil, Trinidad, Costa Rica, Chile, and Venezuela.

Regional and international cooperative schemes

The number of international, regional, and subregional cooperative programs, systems, and networks for the exchange of information is steadily increasing. The trends today are for greater cooperation for a better exchange of information itself as well as an exchange of know-how and experience in managing information systems. Technical, sociopolitical, and cost factors for the development of such systems have been analyzed in the literature.[16] The role of the United Nations and its specialized agencies in establishing cooperative regional and international information systems is generally recognized; these agencies have in fact developed some interesting models of international cooperation in the information field.[17]

Unesco's action in this area aims at strengthening national capabilities for information exchange and creating the necessary mechanisms for sharing experience and resources and for reviewing and coordinating regional activities. Unesco here acts as a catalyst, providing the necessary stimulus, technical backup, organizational methodologies, standards, and tools, as well as limited financial support. Examples of such regional schemes are the Regional Network for the Exchange of Information and Experience in Science and Technology in Asia and the Pacific; the Caribbean Network for the Exchange of Information and Experience in Science and Technology; the Regional Program of Cooperation in the Field of Information in Latin America and the Caribbean; the Documentation Centre for the League of the Arab States; and the Asia-Pacific Information Network for Medicinal and Aromatic Plants.

Member states of the United Nations have expressed their need and wish for a mechanism to facilitate the global flow of scientific and technical information for development across nations. The United Nations Conference on Science and Technology for Development, held in Vienna from 20 to 31 August 1979, specified the functions of such a global information network. Many obstacles have been encountered, ranging from lack of adequate national resources, inadequate national infrastructures, and lack of specialized labor to the absence of international commitment for financing this ambitious idea.[18] So far the valid proposal remains that advocated by the 1971 UNISIST conference: a flexible and loosely connected international network of information systems and services based on voluntary cooperation.

National information policies and infrastructures

Unesco has for many years provided assistance for the creation and development of libraries, archives, and information services, as well as the establishment of units performing a number of special-

16. Jacques Tocatlian, "International Information Systems," *Advances in Librarianship*, 5:1-60 (1975).

17. A. Neelameghan and Jacques Tocatlian, "International Cooperation in Information Systems and Services," *Journal of the American Society for Information Science*, 36(3):153-63 (May 1985).

18. Jacques Tocatlian, "Towards a Global Information Network," *The Use of Information in a Changing World—FID 42nd Congress* (Hague: International Federation for Information and Documentation, 1984), pp. 5-18.

ized information services. In most countries the development of national information infrastructure, such as the national library, the university libraries, the national archives, and many of the sectoral specialized information services, are supported by the public sector. The development of such national information institutions and services is much more coherent and effective if it is done in the framework of an overall national information policy, which represents a commitment of the government to sustain national efforts. In developing countries, where the information industry and private sector in this area are weak, a national information policy is essential.

Unesco has provided practical guidance on policy formulation and planning techniques. The first attempt to render this guidance resulted in the publication in 1974 of *Information Policy Objectives: UNISIST Proposals*,[19] prepared by J. Gray. This publication provided a checklist of 113 possible policy objectives in the field of scientific and technical information. Revised guidelines drawing some of the lessons of experience that can be inferred from past efforts were prepared in 1985 by I. Wesley-Tanaskovic.[20] Unesco recommended to its member states the establishment of a governmental agency to guide, stimulate, and conduct the development of information resources and services and the establishment of a UNISIST/PGI national committee to serve as national links to Unesco's information program. By March 1987 60 member states had designated national focal points and 51 had organized UNISIST national committees. In response to requests from member states, Unesco has assisted in the organization of national seminars on the development of coordinated national information policies and their subsequent implementation through a plan of action. In recent years such national information policy seminars were organized in the following 18 countries: Austria, Botswana, China, the Democratic People's Republic of Korea, Ethiopia, Hungary, Jamaica, Malaysia, Malawi, Mexico, Nepal, Panama, Peru, the Philippines, Sri Lanka, Thailand, Zambia, and Zimbabwe.

Unesco's assistance to the developing countries in establishing and developing information institutions, systems, and services has been quite substantial. The UNISIST/PGI documentation on the subject is abundant, the number of projects, missions, and meetings quite large, and the achievements generally recognized as impressive, taking into account the field to be covered and the relatively modest budget of the program. A few examples should illustrate the variety of activities undertaken.

In the field of archives, Unesco, with the collaboration of the International Council on Archives, has launched the Records and Archives Management Program, under which a very large number of practical guidelines, studies, and manuals have been prepared. In a 1983 listing prepared by F. Evans,[21] 282 such documents and publications were mentioned. National records often relate to the field of science and technology.

In the field of libraries, Unesco, in collaboration with the International Federation of Library Associations and Institutions, assists in the creation and

19. Unesco, *Information Policy Objectives: UNISIST Proposals* (Paris: Unesco, 1974).

20. Unesco, *Guidelines on National Information Policy: Scope, Formulation and Implementation* (Paris: Unesco, 1985).

21. Unesco, *Writings on Archives Published by and with the Assistance of Unesco: A RAMP Study* (Paris: Unesco, 1983).

strengthening of national, university, research, school, and public libraries by providing consultants and organizing pilot projects, congresses, and seminars. A special concern here is an international strategy to ensure universal accessibility to publications. A recent article in the literature provides up-to-date information on the role of Unesco in the development of library services in developing countries.[22]

Another nongovernmental organization with which Unesco cooperates is the International Federation for Information and Documentation. The area of information and documentation is one of particular concern for science and technology. A wealth of guidelines have been produced by Unesco to assist in conducting national inventories of information and documentation facilities and national inventories of current research projects; in evaluating information systems and services; and in establishing specialized services such as referral centers or services for selective dissemination of information. Particular mention might be made regarding cooperation with the Committee on Data for Science and Technology, established by the International Council of Scientific Unions. This cooperation has been fruitful in the area of data sources[23] and data handling in general.[24]

22. Jacques Tocatlian and Aziz Abid, "The Development of Library and Information Services in Developing Countries: Unesco/PGI's Role and Activities," *IFLA Journal*, 12(4):280-85 (Nov. 1986).

23. Unesco and Committee on Data for Science and Technology, *Inventory of Data Sources in Science and Technology—A Preliminary Survey* (Paris: Unesco, 1982).

24. Unesco and Committee on Data for Science and Technology, *Data Handling for Science and Technology* (Amsterdam: North Holland, 1980).

Through Unesco, countries such as China, India, and Morocco are now capable of conducting on-line searches of internationally available data bases. Assistance provided included demonstration, training, equipment, and consultant services. Also through Unesco's assistance, many new information technologies have been introduced in the Third World and many new approaches have been tested and adapted to local needs.

Developing an information work force

Under Unesco's PGI, priority has traditionally been given to activities for the training of both information specialists and information users. Allocations for this subprogram account for 25 percent of the appropriation for the whole program. The kinds of occupation involved in the information field are tending to become increasingly varied. New qualifications are required in data processing and management and in the field or specialized subject with which the information being processed is concerned.

The activities here aim at developing and improving national and regional training programs, preparing teaching materials, and providing training for teachers and refresher courses for specialists. Consultants are sent to countries to advise on the development of training programs and to participate in teaching. Regional postgraduate programs in information science, with substantial support for staff development, have been created and supported in, inter alia, China, Ethiopia, Morocco, Nigeria, the Philippines, Senegal, and Venezuela. Regional seminars in curriculum development are organized; numerous national and regional seminars and refresher courses in a variety of technical subjects are supported; in-service training of teachers and managers, training programs for users, and study tours constitute a major feature of this part of the program.

Numerous publications in this area have been widely distributed and utilized in a variety of subjects such as policy on education, curriculum development, the organization of training courses and their evaluation, the use of audiovisual aids, and the education of users.

Other efforts in Unesco

The preceding pages have attempted to illustrate the main lines of emphasis of Unesco's PGI, which is the main focus for Unesco's action in the area of concern in this article. It should be noted, however, that many other Unesco programs, especially in the Science Sector of the organization, contribute to enhancing the flow of scientific and technical information among nations. In fact, the Science Sector has established a number of specialized information systems concerned with new and renewable energies, hydrology, marine environment, and other matters.

LOOKING AHEAD

In conclusion, it may be said—as we near the 1990s—that the trends will continue to be for increased international cooperation to facilitate the flow of scientific and technical information across national boundaries. Countries around the world have become fully conscious of the importance of scientific and technical information for decision making, problem solving, effective planning, research, and development. There is so much to be gained from a better exchange of information that, in this age of interdependence, it is unlikely that any country can consider itself self-sufficient as far as scientific and technical information is concerned. The principles underlying Unesco's action in this field, the choice of priorities, and the strategies selected have proven to be correct. The fast development of computer and telecommunications technologies and their application in the development of information systems will have a very great impact on this area of human endeavor.

ANNALS, *AAPSS*, **495**, January 1988

The Luxembourg Income Study: The Use of International Telecommunications in Comparative Social Research

By LEE RAINWATER and TIMOTHY M. SMEEDING

ABSTRACT: A computerized telecommunications network, the BITNET-EARN-NETNORTH system, has made it possible for some 588 universities worldwide to efficiently access large-scale statistical data sets stored at one central facility. There are practical and traditional difficulties in comparative international research projects, and the Luxembourg Income Study (LIS) seeks to overcome them. The advantages and disadvantages of housing a statistical data base such as LIS in one place are outlined. Advantages such as building an expert staff who thoroughly understand the data base and securing the privacy and confidentiality guarantees required before nations will grant access to official income statistics are contrasted with the disadvantages of time, cost, and user distance from the data base. The BITNET-EARN-NETNORTH system plus the LIS user package reconciles costs and benefits and allows access by researchers at any BITNET-EARN-NETNORTH site. The challenge for realizing the social research potential of centralized data sets and scholarly colleagueship now lies in creating payment mechanisms and funding consortia based upon principles that can facilitate international research collaboration.

Lee Rainwater is professor of sociology at Harvard University and chairman, Faculty Executive Committee, Harvard Institute of Social Research. His research has most recently been concerned with social stratification from a comparative perspective. He is research director of the Luxembourg Income Study Project.

Timothy M. Smeeding is professor of public policy and economics at Vanderbilt University. His work is in the economics of public policy: the measurement of economic well-being and poverty, the well-being of the elderly, and health care finance. He is project director of the Luxembourg Income Study Project.

THE practical and organizational difficulties of time, distance, and cost have for many years plagued cooperative international research projects in the social and other sciences. While international research societies in all of the social sciences regularly convene meetings, conferences, and congresses at which papers are presented, the papers normally remain parochial in scope, limiting their analyses to the experience of one country concerning a given social issue. The basic physical sciences can most often deal with scientific laboratory problems that transcend social, economic, and political boundaries. It is precisely these boundaries, however, that are usually the essence of comparative social science research because they present the researcher with the opportunity to investigate the ways in which different societies cope with similar social problems.

TRADITIONAL MODE OF COMPARATIVE RESEARCH

Until a short time ago, the most common and pragmatic solution for social scientists interested in comparative research was to call together a group of area experts who would meet once, discuss a proposed project, and then write papers about how their own countries were dealing with substantive issues such as high levels of poverty or prevalent public health problems. The papers would then be assembled into a single volume encompassing an introduction, a variety of chapters referable to individual countries, and a summary paper that attempted, usually in vain, to unify the volume. The results of such endeavors were often perplexing to those who sought to learn from such works or influence policy on the basis of them. In particular, when the country-specific chapters dealt with quantitative data of differential quality, coverage, definition, and scope—which was almost always the case—the results that emerged were frustratingly hard—frequently impossible—to compare across countries. Experiences of the "apples and oranges" variety such as this have for decades discouraged true comparative research in the social sciences.

The changing scene

The rapidly evolving technology of computerized data banks provides a challenging opportunity to assemble multinational data bases that provide a common foundation upon which teams of social scientists can build truly long-term, comparative international research programs. These data bases provide the opportunity to define a range of theoretical and substantive problems and to combine analyses of data from different countries into a single paper or book. This is in fact what the Luxembourg Income Study (LIS) is designed to accomplish.

PLAN OF THE ARTICLE

The next section of the article describes the LIS: its nature and its objectives. Even with the LIS in place, however, important issues of cost, distance, and time must be overcome to create long-term research projects that do not unduly penalize researchers by forcing them to be away from their normal place of work for inordinate periods of time and at high cost. One potential solution to this problem is the dissemination of public-use data sets to individual researchers. Due to respondent privacy and confidentiality problems that usually restrict access to microdata sets, however, this solution is not always possible. Recent advances in statistical methodology combined with greatly increased software capabilities have raised new fears concerning the ability of statistical spies to penetrate even the best confidentiality safeguards. Therefore, the second section of this article discusses telecommunications technology in terms of what it has to offer by way of a solution to these problems, using the LIS as an example.

We describe our use of the BITNET-EARN-NETNORTH interuniversity telecommunications system.[1] We argue that because of the BITNET-EARN-NETNORTH system, the LIS is able to overcome not only distance, time, and cost but also the technological and political or administrative barriers that it faces. Here we find that rather than posing an intrusive threat to individual privacy, modern telescience has offered researchers the opportunity to begin to explore at relatively low cost an entirely new and exciting realm of scientific inquiry that would otherwise be lost.

The final section of the article mentions the long-run implications of such systems for comparative research, both the potential and the difficulties that it affords. We conclude by pointing to the challenge of meeting the short-term and long-term costs of maintaining and expanding both the comparative data base and the telecommunications network upon which it relies as the key elements in fostering future collaborative, cross-national social science research.

DESCRIPTION OF THE LIS

The LIS experiment began in April 1983. Its purpose is to gather in one central location, the Center for Population, Poverty, and Policy Studies (CEPS) in Luxembourg, sophisticated microdata sets that contain comprehensive measures of income and economic well-being for a set of industrialized welfare states. Because of the breadth and flexibility afforded by microdata, each researcher is free to make several choices such as definition of unit—family or household, for example; measure of income; and population to be studied, such as males, females, urban families, or elderly households. These truly comparable microdata create a potentially rich resource for human-resource and related policy research.

As of 1987, the LIS data bank contained data sets from Australia, Britain, Canada, Germany, Israel, Norway, Sweden, Switzerland, and the United States. Data sets from Holland, Denmark, Italy, Finland, France, and Spain will likely be added in 1988. Table 1 gives an overview of these data sets by country, data-set name and size, income year, data sampling frame, and representativeness of the population.

The data base consists of income microdata sets prepared according to a common plan, based on common definitions of income—by source—taxes, and family and household composition and characteristics. Spouses' earnings and average annual wage rates—earnings divided by hours worked—are separately recorded as well. This resource has already proved extremely useful in both basic and applied social and economic research concerned with such human-resource issues as

- —the distribution of household income and the relative income positions of the old and the young, urban and rural residents, and other groups of policy interest, such as single parents;

1. "BITNET" stands for "Because It's Time Network"; "EARN" stands for "European Academic Research Network." NETNORTH spans Canada.

TABLE 1
AN OVERVIEW OF LIS DATA SETS

Country	Data-set Name, Income Year (and size)*	Population Coverage‡	Basis of Household Sampling Frame††
Australia	Income and Housing Survey, 1981-82 (45,000)	97.5§	Decennial census
Canada	Survey of Consumer Finances, 1981 (37,900)	97.5§	Decennial census
Germany	Transfer Survey, 1981† (2800)	91.5**	Electoral register and census
Israel	Family Expenditure Survey, 1979 (2300)	89.0‖	Electoral register
Norway	Norwegian Tax Files, 1979 (10,400)	98.5§	Tax records
Sweden	Swedish Income Distribution Survey, 1981 (9600)	98.0§	Population register
Switzerland	Income and Wealth Survey, 1982 (7036)	95.5‡‡	Electoral register and central register for foreigners
United Kingdom	Family Expenditure Survey,† 1979 (6800)	96.5#	Electoral register
United States	Current Population Survey, 1979 (65,000)	97.5§	Decennial census

SOURCE: Timothy Smeeding et al., "LIS User Guide," LIS-CEPS WP#7 (Center for Population, Poverty, and Policy Studies, Walferdange, Luxembourg, December 1986).

*Data-set size is the number of actual household units surveyed.

†The U.K. and German surveys collect subannual income data, which are normalized to annual income levels.

‡As a percentage of total national population.

§Excludes institutionalized and homeless populations. Northern rural residents—Innuits, Eskimos, Laps, and others—may be undersampled.

‖Excludes rural population—those living in places of 2000 or less—the institutionalized, the homeless, people on kibbutzim, and guest workers.

#Excludes those not on the electoral register, the homeless, and the institutionalized.

**Excludes foreign-born heads of households, the institutionalized, and the homeless.

††The sampling frame indicates the overall base from which the relevant household population sample was drawn. Actual sample may be drawn on a stratified probability basis, for example, by area or age.

‡‡Excludes nonresident foreigners and the institutionalized but includes foreign residents.

—the distribution of earnings for both men and women and their change over the worker's life cycle, including the transition to retirement; and

—comparative studies of the workings of the welfare state and its policies toward the elderly, the disabled, and the unemployed.

A substantive example

In September 1985, a brief set of tabulations and an explanation was presented to a National Academy of Sciences Workshop on Demographic Change and Well-Being of Dependents, held at Woods Hole, Massachusetts. One of the major themes of the workshop was that one can learn a great deal about the relative economic status of U.S. children—under age 18—and the elderly—over age 65—by considering these persons in relation to similar groups in other countries.

Because relative economic status—poverty, low income, and affluence—is dependent upon family situations, we grouped child and elderly data in a way that separated them as much as possible from other parts of the population, constituted of middle-aged adults living with children and adults living alone. An example of the type of data presented to the National Academy of Sciences workshop is shown in Table 2, where relative poverty rates have been separately computed for elderly persons, for children, and for all other adults regardless of their family situations.

Relative poverty is measured here as all persons living in a family unit with adjusted cash after-tax income below half the adjusted median after-tax cash income of the entire population. A simple adult equivalence scale counting the first person as 0.50 adults and all others as 0.25 adults was used. Poverty rates are measured as the percentage of each type of person who meets the poverty definition. Overall poverty rates are also calculated for all persons.

In Table 2, the overall rate of poverty is highest for the United States. The poverty rate for children is much higher for the United States than for any other country, while the U.S. elderly population's poverty rate is third highest. In all countries but Sweden and Canada, adult rates are lower than the rates for children or the elderly.

Research and enrichments under way

The look at U.S. economic status for the National Academy of Sciences workshop is but one instance of the uses made of the LIS data base to study income, poverty, the relative economic status of one-parent families, of children, and of the elderly, and the overall distribution of government cash transfers versus direct taxes.[2] In addition, projects to add noncash income and to explore the role of women's earnings in family income are currently in progress.

The LIS has now moved beyond the initial experimental stage to provide a data bank that can be perpetually updated and expanded to include the most recent data available for any and all nations that have high-quality income microdata sets and that choose to participate. The data sets will be updated during 1988, adding 1984-85 cross-section data sets and the initial waves from several new European household panel studies.

Research obligations and responsibilities

The LIS Project and data set are permanently housed at the CEPS Research Center in Luxembourg. The data

2. Further information about LIS and additional examples of the uses of its data base can be found in working papers, available from CEPS, Cast Postale 65, L-7201 Walferdange, Luxembourg: "An Introduction to LIS—The Luxembourg Income Study"; "Poverty in Major Industrialized Countries"; "Income Distribution and Redistribution"; "Age and Income in Contemporary Society"; "Comparative Economic Status of the Retired and Nonretired Elderly"; "Relative Economic Status of One-Parent Families"; "LIS User Guide"; "Economic Status of the Young and Old in Six Countries"; "An International Perspective on the Income and Poverty Status of the U.S. Aged: Lessons from the Luxembourg Income Study and the International Database on Aging."

TABLE 2
RELATIVE POVERTY RATES AMONG ELDERLY PERSONS, CHILDREN, AND ADULTS (Percentage)

Country	Overall Rate	Elderly Rate	Children's Rate	Adults' Rate
Canada	12.1	10.3	16.8	10.6
Germany	6.0	9.3	6.3	5.7
Israel	14.5	22.0	18.6	10.4
Norway	4.8	4.6	5.6	4.1
Sweden	5.0	0.0	5.2	6.7
United Kingdom	8.8	20.8	10.4	5.4
United States	16.9	17.4	24.1	12.9
Simple mean	9.7	12.5	12.4	8.0

SOURCE: Timothy M. Smeeding, "Relative Poverty Rates among Children and Elderly in Seven Nations" (Paper delivered at the National Academy of Sciences Workshop on Demographic Change and Well-Being of Dependents, Woods Hole, MA, 4-5 Sept. 1985).

NOTE: The relative poverty rate is the number of persons of each type as a percentage of each type of person with family incomes, adjusted for equivalence, less than one-half median family income, adjusted for equivalence.

*People were counted as elderly if they were age 65 and older.

†People were counted as children if they were below the age of 18. Children's income is measured by that of their parents.

are stored on the government of Luxembourg's central computers, which are accessed via several computer terminals at CEPS under the strict rules of the government of Luxembourg's data access and privacy laws.

Once research papers or reports are prepared from the LIS, the researcher is required to make the results available as a LIS-CEPS Working Paper. In this way we can document previous LIS research from those interested in furthering the use of our network and we can provide for a statistical review of results by LIS member-country central statistical offices. While there is no charge for reasonable use of the LIS data by member countries who have joined financial forces to underwrite the maintenance and renewal of the data base, minimal user charges must be levied on researchers from nonmember countries and international research organizations to pay for data preparation: programmer salaries, data-set computer maintenance charges, and transaction costs. Cost estimates depend on expected use and the difficulty or ease of proposed manipulations.

ACCESS TO THE LIS

In order to use the LIS data base, a system of communication between the researcher and the data set is a necessity. Either the researcher must travel to the data, or the data must be transported to the user. Excluding the possibility of computerized telecommunications, researchers could access the LIS data base in Luxembourg either by traveling to Luxembourg or by using traditional telephone and mail services. These alternatives are both costly and time consuming, especially for transoceanic access to the data base. Research-funding organizations are extremely suspicious of international travel due to its high cost and the supposed personal-consumption—that is, tourist—flavor of such endeavors. Moreover, such travel is disruptive to the researcher, forcing interference with normal job duties and home life and requiring acclimation to work in a foreign environment. On the other hand, there are advantages to on-site access, in particular the ability to interact with the expert staff who thoroughly understand the data base, including its nuances and idiosyncrasies. Because there is usually no perfect substitute for interactive face-to-face discourse with such experts, especially when first using a data set, remote access systems face the challenge of developing user-friendly modes of discourse as an alternative to this verbal interaction.

The usual and preferred alternative to travel is for the data center to create a public-use—that is, completely unrestricted—data file that can be exported to the researcher at minimal cost. Because of the strict privacy restrictions and confidentiality assurances under which some foreign central statistical offices have loaned the LIS copies of their data sets, however, LIS public-use tapes are at present impossible to provide. Despite the incredulity of American researchers who are accustomed to public-use files in their own country and consequently fail to understand why they cannot be created in the case of the LIS, foreign governments can make a strong case for data access limitations based on political concerns about confidentiality threats. Censuses of the population scheduled for 1981 in the Netherlands and 1983 in West Germany had to be postponed due to public concern about privacy, confidentiality, and access to data. Similar privacy issues in Sweden have led to severe criticism of Statistics Sweden and, much to the dismay of all persons involved, a substantial increase in refusal to be included in Swedish income and labor force participation studies. Due to such concerns as these, the restrictions placed on access to the LIS data files are severe. Direct access is restricted to the staff of the LIS project center in Luxembourg under the supervision of Gunther Schmaus and Brigitte Buhmann, the LIS technical team, who have sworn to uphold the Luxembourg government privacy restrictions.

In summary, the dual problems of limited direct access due to cost and distance, and restrictions on secondary public-use distribution of the data sets have created a severe technical problem for the LIS. Fortunately, a solution to both problems becomes possible due to advances in computer networking, which ties geographically distant researchers to the centrally located data base and its technical support staff. Although there are telecommunications alternatives to BITNET-EARN-NETNORTH—such as long-distance dial-up to a time-sharing system or use of commercial public packet-switched networks—low line-leasing costs and the absence of store and forward fees to individual users make BITNET-EARN-NETNORTH the key to the operational success of the LIS.[3]

BITNET-EARN-NETNORTH AND THE LIS

The BITNET-EARN-NETNORTH system is an electronic mail and file transfer network available at some 588 academic and research centers in the United States, Japan, Canada, Great Britain, Europe, Scandinavia, and the Middle East. The diffusion of this network has been extremely rapid. By the

3. Further information on the network is available from the BITNET Network Information Center at EDUCOM, 777 Alexander Road, Princeton, NJ 08540, telephone (609) 520-3377.

turn of the decade, virtually all major universities and social science research centers should be connected to the system.[4] This rapid spread is due to two basic features: low cost and ease of use. A LIS user with a BITNET address—log-on name and computer node—can type in a message—a letter, program, or paper, for instance—which is then transferred by leased lines from university to university host computer until it reaches its final destination in the United States, or to a link, which connects to the EARN network via satellite. The process then repeats to complete the route within Europe, ending up at the Luxembourg Computer Center. The message is held at the center until it is retrieved by a LIS staff member. Once the message is received, the Luxembourg staff either replies to the message or sends a data request to the central computer on which the LIS is stored.

Accessing procedures

In order to facilitate the procedure of accessing LIS data, two separate and independent steps are necessary. First, the user, or potential LIS researcher, must have enough information to be able to request efficiently the output that he or she desires. This requires a complete and user-friendly package to serve as an introduction to the LIS. Second, the LIS staff must be able to process the request and return the output to the user quickly and efficiently, being sure to protect the confidentiality of the file.

4. For a good overview of the network, see Daniel J. Oberst and Sheldon B. Smith, "BITNET: Past, Present, and Future," *EDUCOM Bulletin*, 21:10-17 (Summer 1986).

Once a potential LIS user has contacted the project, the next step is to send the person a brief document that describes the LIS and its mode of functioning. Researchers who wish to proceed in obtaining data may then request the complete user package from the LIS. This package includes

—a technical description of each country data file that goes into the LIS, including sampling frame, expected sampling and nonsampling errors, and other pertinent information;
—a list of definitions of variables, which explains in detail the exact income components from the raw country data file that went into each LIS variable. For demographic variables, this explanation includes the exact wording and codes for such other variables as occupation, education, and marital status in the case of each country. For LIS income variables, the maximum and minimum values, mean, median, and percentage of population receiving each type of income are included along with the name of the income components from the country data set that have been included in the LIS variable;
—an institutional information codebook, which includes a basic description of those income components that are social transfer programs: history, overall outlays, eligibility rules, and bibliographic sources for additional information on each such income source in each country;
—a list of standard recodes of LIS income definitions—for example, pretax income and disposable income—and other recodes—for instance, marital status; one-parent families—for those who wish to

compare their results to earlier LIS analyses using these same concepts;

—a sample data file containing a random sample of about 200 records from each country. This sample is used to test data runs to ensure computer software commands and correct specifications; and

—a package of technical request information, including available software packages and EARN-BITNET technical conventions for sending requests.

A fee of about U.S.$20.00 is paid for this package. The package provides the potential user with answers to most questions that one could initially ask about the LIS. Armed with such a package, it is relatively easy to ensure that timely, nonduplicative, and nonwasteful output requests are sent and returned to the LIS data center with a minimum of turnaround delay. Moreover, this package substantially reduces the amount of up-front investment that the researcher needs to make in order both to understand and to access the data file. Given test output from the sample data file, the researcher will have the wherewithal to debug the software used to request the data and will also have some idea of the sensibility and utility of the output itself.

The dissemination process at CEPS

Once the request, or job, is sent via BITNET-EARN-NETNORTH and received in Luxembourg, specially designed software reads through it, verifies its consistency, and sends it to the main computer in Luxembourg. At present, job requests can only be processed using the SPSSX software package. Eventually, SAS, LIMDEP, and other widely used social science packages will also be available. The finished output is returned from the main computer to the LIS center, where it undergoes data and confidentiality protection review via software that checks that raw data is not being transmitted. Then the output file is sent back to the researcher using the BITNET-EARN-NETNORTH network.

The two key portals of request submission and output retrieval are under the control of the LIS center staff only. Distant researchers can neither directly access the data set via job submission nor directly receive output with positive action on their part. In this way, job input and output can be screened to prevent the violation of data protection and confidentiality laws. While turnaround is not instantaneous, we hope that overnight job submission and output return will become the norm in cases where researchers realize not only their local time at the time they submit the job, but also the local time at which the message will reach the LIS center, which is six to nine hours later than the local time in the United States.

While systems somewhat similar to the LIS are functioning within the United States, such as the Survey of Income and Program Participation (SIPP) Access Center at the University of Wisconsin at Madison, the LIS system is unique due to its virtual worldwide accessibility. Other systems, such as the SIPP Access system, are in place mainly to facilitate user understanding of a very complex, longitudinal data file such as SIPP, for which public-user files are available but not usable without considerable intermediary assistance. While the SIPP Access Center does allow the user an opportunity to be introduced to the data set more easily, user friendliness is its primary advantage. Maintenance of privacy and confidentiality, and overcoming the costs of distance are the added features of the LIS EARN-BITNET access system.

LONG-RUN IMPLICATIONS

Just as office and home are replacing central, on-campus computer facilities as the sites of research and administrative activities, in the long run it may prove more economical to work with data at geographically remote locations than for each investigator to start from scratch by installing raw data at his or her home base—at least with large, complex data bases, as the SIPP Access Center experiment has shown.

As the LIS is beginning to demonstrate, however, remote telecommunications access coupled with input and output screens possesses the additional advantage of providing the data base operators with a means of assuring protection of privacy and confidentiality to the suppliers of the input data sets. Without these protections, there would be no LIS because there would not be the input data sets that are its essence.

While the uniqueness of the LIS data base and of EARN-BITNET-NETNORTH are the strengths of our project, however, they are also its weaknesses. Comparative social policy research is in its infancy. In its second year of operation, the LIS is at approximately the same stage of development that longitudinal household panel data research was at in 1969, the second year of the Panel Study of Income Dynamics. While this panel study has gone on to create fresh and exciting opportunities and methodologies for social science research, it took a decade or more for the tools, strengths, and weaknesses of panel data analysis to permeate the social sciences sufficiently to really catch on. Similarly, experience with empirical comparative social research is now a virtual unknown for most social scientists. The enormity of the enterprise of learning to think cross-nationally is one that few have bothered to undertake and even fewer have mastered. The LIS collaborators struggle with this challenge almost daily. Regardless of the user friendliness of the technical process for accessing the LIS data base, meaningful interpretation of the results is something that requires—as do most highly rewarding life endeavors—time, effort, and commitment on the part of the researcher. In this sense, the real job of using the LIS is just beginning.

Another, more immediate concern is that of ensuring funding mechanisms that would provide the basic public goods that the LIS requires: maintenance and upkeep of both the central LIS data base and the EARN-BITNET telecommunications system. By early 1988 we expect that at a reasonable annual cost, the initial nine LIS countries will have joined in a cooperative funding consortium that underwrites the basic cost of LIS data-file maintenance and renewal for five years. This funding will permit reasonable usage of the LIS at no cost to researchers in all nine countries.

As of January 1987, U.S. institutions have been paying an annual fee to belong to BITNET. The fee ranges from $750 to $8000 and is proportional to annual operating budget. There is no charge for message transmission. The combination of low cost for both data and telecommunications linkage will allow LIS users to achieve maximum economies of scale in the sharing of distributed data bases.

Under similar regimes, we can expect

international collaborative research projects and colleagueship to reach their full potential. Under alternative higher-cost regimes, especially for telecommunications, we are less sanguine about comparative social research reaching its full potential.

ANNALS, *AAPSS*, **495**, January 1988

The World Climate Program: Collaboration and Communication on a Global Scale

By EUGENE W. BIERLY

ABSTRACT: This article discusses the rationale and history of the World Climate Program (WCP) as a prime example of gains in scientific knowledge achievable only through collaboration and communication on a worldwide basis. The WCP is managed jointly by the World Meteorological Organization and the United Nations Environmental Program, both of which are specialized agencies of the United Nations, and by the International Council of Scientific Unions. This unique arrangement has both given strength and presented problems in getting governments and scientists from all over the world to work together in the pursuit of program goals. Vital to this work are the tools made available by contemporary communications technology, particularly supercomputers and satellites. Nevertheless, the availability and usefulness of those tools does not supplant the more basic groundwork that has to be laid and maintained in order to conduct global research. The necessary groundwork requires intra- and intergovernmental collaboration, and also continued progress in the underlying science base. The WCP is composed of the World Climate Data Program, the World Climate Applications Program, the World Climate Impact Program, and the World Climate Research Program.

Eugene W. Bierly is director of the Division of Atmospheric Sciences, Directorate for Geosciences, National Science Foundation. He received the Ph.D. and M.S. degrees from the University of Michigan in meteorology. He is a fellow and past president of the American Meteorological Society and was a member of the American Political Science Association (APSA) while holding an APSA Fellowship in Congressional Operations. He has published numerous articles on atmospheric sciences and science policy.

THE World Climate Program (WCP) represents many elements of a science program that is composed primarily of the physical sciences but is composed also of elements from the social and biological sciences. Progress is being made in the WCP because of cooperation between nearly a dozen U.S. government agencies and several agencies of the United Nations. These represent the governmental input to the program. The nongovernmental input comes via the National Academy of Sciences, within the United States, and the International Council of Scientific Unions (ICSU), representing other academies of science around the world. The WCP already can point to some successes; however, its ultimate success will be determined only if the science base needed to answer many of the issues that confront human beings and their environment is used carefully by decision makers so that legal circumstances and administrative decisions are in concert with scientific results.

BACKGROUND AND HISTORY OF THE WCP

The WCP is a natural outgrowth of the Global Atmospheric Research Program (GARP), whose original planning began in the late 1950s.[1] GARP was designed to study the physical processes in the troposphere[2] and in the stratosphere[3] that are essential for an understanding of (1) the transient behavior of the atmosphere as manifested in large-scale fluctuations that control changes of the weather[4]; understanding this behavior could lead to increasing the accuracy of forecasting over periods from one to several weeks; and (2) the factors that determine the statistical properties of the general circulation of the atmosphere; understanding these factors would lead to better understanding of the physical basis of climate.[5] The GARP field programs that were designed to provide the data required for the design of theoretical models and the testing of their validity now have been completed; however, analyses are proceeding still and will continue for some years.[6]

1. Jule G. Charney, *The Feasibility of Global Observation and Analysis Experiment*, A Report of the Panel on International Meteorological Cooperation to the Committee on Atmospheric Sciences, National Academy of Sciences-National Research Council, 1965. This report is reprinted in *Bulletin of the American Meteorological Society*, 47(3):200-220 (Mar. 1966).

2. The troposphere is the lower portion of the Earth's atmosphere extending from the surface to 10 to 20 kilometers. The troposphere is characterized by decreasing temperature with height, appreciable vertical wind motion, appreciable water vapor content, and weather.

3. The stratosphere is the layer of the atmosphere above the troposphere extending to 20 to 25 kilometers. The stratosphere is characterized by stability and persistence of circulation patterns. Ozone concentrations are highest in this layer.

4. Weather consists of short-term—minutes to weeks—variations of the atmosphere, popularly thought of in terms of temperature, humidity, precipitation, cloudiness, visibility, and wind.

5. World Meteorological Organization and International Council of Scientific Unions, Report of the Study Conference on the Global Atmospheric Research Program, jointly sponsored with the International Union of Geodesy and Geophysics, Geneva, 1967. Climate comprises the long-term—ranging from months to eons—manifestations of weather. The climate of a specified area is represented by the statistical collective of its weather conditions during a specified interval of time.

6. Jay S. Fein, Pamela L. Stephens, and Kristyne S. Loughran, "The Global Atmospheric Research Program: 1979-1982," *Reviews of Geophysics and Space Physics*, 21(8):1076-96 (June 1983).

The rise of public concern

The second objective of the original GARP concept concerning climate did not receive much attention immediately. In later years, during the early 1970s, however, climate became an important public issue. The early 1970s witnessed a number of climatic events that had disastrous consequences. The Sahel region of Africa suffered a five-year drought that brought famine and death on a large scale. In 1972 the Soviet Union had a significant drought and had to buy grain from the United States. That same year there was an El Niño, a warm current off the coast of Peru, that destroyed the anchoveta fishery. The year 1974 brought a severe monsoon to India that reduced food production. Cold weather reduced Brazil's coffee crop in 1975. Cold weather in Europe in 1976 caused widespread economic dislocation. An unusually cold winter in the United States the same year caused many industrial and school closings. Thus the attention of the public was drawn to the extremes of climatic variability.

U.S. government reactions

Was there any reaction to these disasters? There was indeed. Several agencies of the U.S. government established programs to study the dynamics of climate. The Climate Dynamics Program established in 1974 by the National Science Foundation is one result of that reaction. The U.S. Congress passed the National Climate Program Act of 1978, which set up an interagency office to coordinate the U.S. government's activities and to begin new ones where needed.[7] This office was established within the National Oceanic and Atmospheric Administration, and it was responsible for reporting to the Congress via the administrator of the National Oceanic and Atmospheric Administration, the secretary of commerce, and the president. The National Climate Program Office was responsible for writing the initial five-year plan laying out programs that were needed and keeping them within the framework of ongoing budgets of various agencies that were involved. There have been eight reports sent to Congress since 1979 describing the ongoing program and citing any needs, administrative or budgetary, that might be felt.[8]

International reaction

Meanwhile, activities were taking place internationally, under the auspices of the World Meteorological Organization (WMO). The Joint Organizing Committee for GARP, a committee composed of 12 scientists selected jointly by WMO and ICSU to give guidance to GARP, realized as early as 1973 that the second objective of GARP deserved some attention. This concept was not readily acted on by most of the scientific community because a great deal of effort was being expended in those days on the implementation of the major GARP experiment, the Global Weather Experiment, held in 1978 and 1979. Nevertheless, the Joint Organizing Committee proceeded to call the Study Conference on Climate, which was held in Stockholm, Sweden, in July of 1974. This study conference set the stage for all later activities under the

7. Public Law No. 95-376, 95th Cong., 2d sess. (17 Sept. 1978), 92 Stat. 601-5.

8. *Annual Report National Climate Program* (Rockville, MD: U.S., Department of Commerce, National Oceanic and Atmospheric Administration, National Climate Program Office, 1979-86).

ICSU-WMO banner.[9] The conference was supported also by the United Nations Environmental Program.

The Joint Organizing Committee continued to discuss climate at its meetings and in fact identified early on two important components of a climate program that deserved attention. Those components were ocean dynamics—that is, the interaction between the atmosphere and the oceans' surface—and the effect of clouds on long-wave and short-wave radiation. Both of these topics were discussed extensively, and plans were laid to identify how these components could be studied so that the results would be useful to modeling the general circulation of the atmosphere successfully. From the very beginning the WMO was envisioned as the scientific leader for the climate program. This was so primarily because WMO had a broad cross-section of member countries, it had the key to governmental action as evidenced by GARP, and, through its connection with ICSU, it also had a way to call on the academicians of the world.

By 1977 the Executive Committee of the WMO decided to request the Secretary-General of the WMO to make specific proposals regarding the establishment of the WCP within the WMO. In addition the Executive Committee decided to convene a high-level scientific and technical World Climate Conference in early 1979 to be attended by physical and social scientists as well as experts from climate-sensitive branches of national economies, including agriculture, energy, water resources, fisheries, and health. The main purposes of the conference were to review knowledge of climatic change and variability, due to both natural and anthropogenic causes, and to assess possible future climatic changes and variability and their implications for human activities. There were more than 350 attendees at the conference. The assembled group heard papers from users of climate information, researchers, politicians, and gatherers of climate data. It was a diversified group, many of whom had had interests in climate studies for years and others who were just beginning to understand that climate played an important role in the life and welfare of human beings.[10]

The World Climate Conference outlined, in general, the problems of climate studies and did something even more important. It discussed some of the interfaces between climate and human activities. These areas could be referred to as impact studies, depending upon whether there was an impact on policy considerations or, at least, on applications of climate information to selected problems. The major interfaces were with regard to climate's relation to food, water, energy, and urban planning. Although many important issues were discussed at the conference, it was May 1979, when the WMO's Eighth Congress met, before the WCP really came into being.

Components of the WCP

Eventually, the WCP was structured to have four components: the World Climate Data Program (WCDP), the World Climate Applications Program (WCAP), the World Climate Impact Program (WCIP), and the World Cli-

9. *The Physical Basis of Climate and Climate Modelling,* Report of the International Study Conference in Stockholm GARP Publications Series no. 16 (Geneva: World Meteorological Organization, 1975).

10. *Proceedings of the World Climate Conference,* WMO—No. 537 (Geneva: World Meteorological Organization, 1979).

mate Research Program (WCRP). One of the unique facets of the WCP is that the responsibility for managing its several components was not all vested in the WMO. The first two areas became the responsibility of the WMO because the WMO is the international agency responsible for observing, collecting, and disseminating meteorological data. Those data serve as the basis for weather forecasting, but they also serve as the basis for climate studies. In addition, the WMO was given the responsibility to direct, oversee, and coordinate the entire WCP.

Responsibility for the WCIP was given to the United Nations Environmental Program. The WCRP responsibility was assigned to WMO in cooperation with ICSU.

An international governmental-nongovernmental alliance

The alliance between WMO, an intergovernmental group, and ICSU, a nongovernmental group, had worked successfully during GARP, so it was thought that such an alliance could work again for climate studies even though it was recognized that studies of climate would be much more difficult and complicated than GARP had been. Such an alliance brings together the funding, planning, and other attributes of governments and the intellectual power of the world's academic community. Thus scientists anywhere in the world, whether in government or not, had a mechanism that would allow them to work together.

The actual working mechanism that allows WMO and ICSU to work together is the Joint Scientific Committee (JSC), composed of about 12 well-recognized scientists who were selected from throughout the world as individual scientists and not as representatives of their governments. The JSC is the equivalent of the Joint Organizing Committee for GARP. Its responsibility is to lay out in general terms plans for the WCRP. The JSC meets about every nine months to discuss how present programs are proceeding and what future ones should be planned. If a new program is selected, then the JSC gives general guidelines on how that program is to be implemented. The reports of JSC meetings are disseminated widely so the community of interested parties can be aware of the JSC's deliberations and plans.

Oceanographic participation

Because oceanography is such an important part of the WCRP, the oceanographic community established a Coordinating Committee on the Climate of the Oceans to plan the oceanographic aspects of the WCRP. The committee is similar to the JSC in that it is composed of scientists from the Scientific Committee for Oceanographic Research of ICSU and the International Oceanographic Commission of the United Nations Educational, Scientific, and Cultural Organization. Thus it has the ability to blend governmental views from the International Oceanographic Commission and academicians' views from the Scientific Committee for Oceanographic Research. It works closely with the JSC due to close staff liaison and sometimes overlapping membership on the two committees.

THE WORLD CLIMATE DATA PROGRAM

The WCDP is truly the foundation of the WCP. All of the aspects of the WCP are dependent on the availability of relevant climate data. In order to manage, plan, and carry out programs within the WCP, it is necessary to have long time-series of the necessary data with sufficient spatial coverage. Thus the purpose of the WCDP is to ensure reliable climate data from the atmosphere, oceans, cryosphere, and land surfaces including the biosphere. These data have to be easily accessible and exchangeable in an acceptable format.

The long-term objectives of the WCDP are to

—improve national systems for climate data management and the availability of referral information on station networks, data sets, and publications;
—coordinate existing data exchange systems and consolidate requirements for observations and data exchange;
—assist nations in building data banks to serve their needs; and
—develop a monitoring, diagnostic, and dissemination system to highlight climatic events that may effect activities of human beings.

The following is a description of projects that are part of the WCDP.[11]

Projects within the WCDP

Improvement of climate data management systems and user services. Data are being rescued by microfilming original manuscripts before they deteriorate. Technical guidelines are being prepared on observing networks, data quality control, data processing and management, organization of data banks, and user services. Reference climatological stations will be established, and data sets, inventories, and catalogs of existing data will be compiled. Computerized procedures will be initiated at centers. Education and training workshops plus seminars will be held to aid in molding the data requirements to national needs. Coordination between countries and regions will be promoted.

Transfer of technology in climate data management and user services. The climate computer (CLICOM) project has as its goal the transfer of technology in climate data management and user services through the provision of comprehensive specifications for microcomputer systems. The CLICOM is a package concept that includes computer hardware, user-friendly software, and training. Within ten years it is hoped that all meteorological services throughout the world that desire such equipment will have a CLICOM system. An important component involves the application of climatic data to problems in agriculture and water resources management.

Consolidation of climate data requirements and improvement of data exchange. Climate applications and monitoring require time series and the operational exchange of monthly and daily data that have sufficient spatial density. A composite observing system using satellite remote sensing and a surface-based network is planned. It will be necessary to increase the exchange of data from one to ten stations per 250,000 square kilometers, and, as requirements change, more data and parameters may need to be observed and exchanged. A reference climatological station network will be

11. The description of WCDP projects and of projects of the other WCP component programs later in this article are based on World Meteorological Organization, "The World Climate Program 1988-1997," *Second WMO Long-Term Plan*, pt. 2, vol. 2 (Geneva: World Meteorological Organization, 1987).

established, and the exchange of daily precipitation data will be improved using the Global Telecommunications System.

Implementation of the climate data referral system. There is a need for concise information on climate data. Specifically, information is needed on the availability of data sets, data summaries, and station networks. Initial catalogs have been published, as have input and output formats and access codes for computerized retrieval and storage of information. Information on data centers and their holdings are included as well as cross-references to WMO, ICSU, International Oceanographic Commission, and United Nations Environmental Program centers.

Development of global and regional climate system data sets. Global data sets must be available in order to pursue climate diagnostic studies and climate prediction research. For the WCP, a variety of data sets will be required comprising regular observations of surface and upper-air meteorological variables—including remote-sensed parameters—as well as marine, oceanographic, cryospheric, land-surface and subsurface, vegetation, soil, topography, and other variables. Use of image processing, enhancement, and overlay techniques will aid studies.

Development of a climate system monitoring capability. The capability to monitor climate systems is needed to provide nations with information on large-scale climate system fluctuations and to facilitate the interpretation of anomalous climatic events. These aims are accomplished through monthly bulletins, special advisories, and annual summaries. Data and processed information for the climate system monitoring come from many sources in several countries such as the Climate Analysis Center of the National Oceanic and Atmospheric Administration within the United States, the University of East Anglia, in the United Kingdom, the Bureau of Meteorology in Australia, and the Hydrometeorological Service of the USSR.

THE WORLD CLIMATE APPLICATIONS PROGRAM

The real payoff to nations is the WCAP, for it is from this program that the uses for climate information will be identified and applied to the many problems that exist around the world. The WCAP has four subprograms: WCAP-Food, WCAP-Water, WCAP-Energy, and WCAP-Other Applications. Many of the activities within the WCAP are conducted with U.N. agencies such as the Food and Agricultural Organization and the United Nations Educational, Scientific, and Cultural Organization, and with other components of the WMO.

Projects within the WCAP

Definition of user information requirements for specific climate applications. It is anticipated that by 1995, 95 percent of the nations of the world will have defined their requirements for major climate applications.

Description of climate effects on food production. Past, present, and future purpose-oriented forecasts will be used to determine the influence of climate on the productivity of specific crops, forests, pastures, and animals.

Determination of climate implications for water resources management. A better understanding of the impact of climate, climate variability, and climate change on water resources will be developed.

Determination of climate implications for energy. What information is useful, how it will be collected, and how it will be presented will be determined.

Implementation of climate applications in other human activities. The effects of climate on buildings and human settlements will be documented in a report called Urban and Building Climatology. The specification of user requirements, creation of data formats, making of special observations, production of guidelines, and development of computer data bases will be undertaken.

Assistance in employing existing practical methodology. National climate application programs will be developed, and information on economic benefits of climate applications will be exchanged.

Combating effects of drought. Studies of observational network density, preparation of drought probability maps and guidelines on the use of climate data to combat the effects of drought, and studies on the assessment of semiarid zones for the support of human activities will be undertaken.

Development of a Climate Applications Referral System (CARS). The upgrading and completion of CARS-Food, the completion of CARS-Solar and Wind Energy, and the beginning of CARS-Water, Urban and Building Climatology, and Climate and Human Health will be accomplished.

Promotion of the development of new climate application methods. The use of satellite information will be promoted. New ways to present data and information and new methods to use statistical and real-time data for the formulation of application-oriented forecasts will be developed.

THE WORLD CLIMATE IMPACT PROGRAM

The purpose of the WCIP is to bring climate information into the consideration of policy alternatives and to warn of economic, political, and social impacts that climate change and variations might bring forth. This program is very sensitive because expectations are high, yet what can be delivered credibly is much less. Thus there is frustration with the scientific community because important questions cannot be answered. There also is frustration on the part of the scientific community itself because of the impatience of users and the knowledge by scientists that the answers desired will take years before they are available, if ever.

Projects within the WCIP

Assessment of the role of CO_2 and other radiatively active gases in climate variations and their impact. The assessment of effects on the atmosphere of changing greenhouse gas[12] concentrations in the atmosphere and the effects of the resulting climate change on human beings and their environment will be undertaken about every five years, with particular emphasis on the socioeconomic aspects.

Dissemination of information on the greenhouse gas/climate change issue. Through guidelines, brochures, and audiovisual materials, nations and individuals will be sensitized to the issue of greenhouse gas and climate change.

Regional assessment of impacts of

12. Greenhouse gas is a gas that traps solar energy in the atmosphere, thus contributing to the warming of the Earth. Examples of such gas are carbon dioxide, methane, nitrous oxide, and chlorofluorocarbons.

climate change. Six regional assessments will be carried out, three in developed regions and three in developing regions, to aid in national decision making.

Assessment of the impact of sea-level change. Analysis of the impact of sea-level change on environmental sectors and on human socioeconomic sectors will be undertaken, consistent with greenhouse-gas-induced warming of coastal, estuarine, and river delta regions.

Advisory group on greenhouse gases. The determination of the need for guidelines for a global framework convention for the protection of the tropospheric climate will be pursued.

Dissemination of knowledge and methods of climate impact assessment. Handbooks on climate impact assessment will be developed, and demonstration projects will be carried out.

Study of impacts of drought in developing countries. Support will be given to the African Centre of Meteorological Applications for Development. Support will also be devoted to research on the socioeconomic impact of droughts and to the development of strategies for the prevention, mitigation, and avoidance of adverse impacts of drought.

Monitoring of the impacts of the El Niño Southern Oscillation events and other teleconnections. Information will be exchanged on the impacts of climatic events on climatically sensitive sectors such as agriculture, fisheries, and water resources.

International network of climate impact studies. National and regional climate impact studies will be identified and national climate programs invited to exchange information. An inventory of climate impact studies will be made. A bibliography and a directory of scientists and institutions will be published.

Establishment of national impact studies programs. The development of national climate programs that have as a component climate impact studies will be encouraged. Once developed, they will become a part of the climate impact studies network.

Support to national climate impact studies. Special climate impact studies will be undertaken when gaps in knowledge are identified or when sufficient resources are not available in a region where such studies are needed.

THE WORLD CLIMATE RESEARCH PROGRAM

The research component of the WCP has as its goal the improvement of our knowledge of climate, climate variations, and the mechanisms that might bring about climate change so one can determine the extent to which climate can be predicted and the extent of human influence on climate. Such a program consists of studies of the global atmosphere, oceans, sea and land ice, and the land surface. The development of models to simulate the climate system so that the sensitivity of climate to natural and man-made influences can be determined is an important aspect of this work.

Projects within the WCRP

Global climate analysis and model development. Progressive improvement of the formulation of all significant physical processes in the climate system will be undertaken through numerical experiments and comparison with global climatological data or detailed observations during intensive field studies of specific processes. Global estimates of derived quantities such as surface fluxes of momentum, energy, and water will be produced through three- and four-dimensional analyses of the primary meteorological fields.

Research on climate processes. A refining of the formulation, in terms of climate model parameters, of the physical processes that are significant in determining the mean state of climate and its variations is needed. Cloud-radiation feedback, atmospheric boundary layer exchanges, and hydrological processes over the land surface are priority areas for all aspects of climate research and for long-range weather prediction on time scales of one to two months. A fully interactive treatment of sea-ice processes in the polar oceans and studies of polar ice sheets and glaciers also are needed.

Study of the tropical ocean and global atmosphere (TOGA). Prediction of the evolution of the coupled system comprising the tropical oceans and the overlying atmosphere is important to the study of the mechanisms that determine the interseasonal and interannual variability of global large-scale monsoon flow.

World ocean circulation experiment (WOCE). Understanding the world ocean circulation and its relation to climate is necessary. Emphasis will be on the large-scale average heat and fresh water fluxes and their annual and interannual variations, the variations of the space-averaged ocean circulation on a time scale of months to years and the statistics of smaller-scale motions, and the volume and location of water masses with a ventilation time scale of 10 to 100 years.

Study of climate forcings. Determining the sensitivity of climate and climate variations to possible causal factors such as changes in solar radiation, composition or particulate matter loading of the atmosphere, land vegetation, and other earth environmental or external factors will be undertaken.

Study of global change. How the earth's land, sea, and atmosphere interact through the combination of physical, chemical, and biological processes and how ecosystems function to absorb, buffer, or generate changes on the global scale are important aspects of global change. An extension of the quantitative modeling methods used in earlier studies will be used to incorporate global or large-scale ecosystems using suitable parametric representations. Ecosystems will be characterized by expressing fluxes of the energy and chemicals that they absorb or yield, in quantitative terms, based on global surveys of large ecosystems using fast data acquisition techniques such as satellite remote sensing.

CONCLUSION

The WCP has addressed some important communications problems in very unique ways. It has brought in the U.S. scientific community through the National Academy of Sciences and has allowed the various federal agencies to work together through several interagency committees. Internationally, it has built upon an earlier relationship between WMO and ICSU, extended it to the climate research area, and continues to bring together the world's academicians and government scientists to make it work.

The WCP is grounded upon sound scientific principles and a successful set of global-scale experiments that were carried out under the aegis of GARP. The WCP is a logical extension of this work. It incorporates much more oceanography into the research because the time scale of the predictions is months to years, and on those time scales, the oceans play a major role in storing, transporting, and distributing the energy received from the sun at the Earth's surface.

The JSC, jointly sponsored and supported by ICSU and WMO, is the mechanism that lays out the scientific program for the WCRP. Members of the JSC are selected on the basis of their scientific credentials and not on political considerations. Nevertheless, there is a balance maintained on the JSC between the superpowers and other countries that can contribute to the scientific efforts of the WCRP.

Credible data drive the WCP. Accurate and timely observations, made at the proper places, disseminated through the WMO, and archived in an acceptable format are now available. This achievement is due to the long history of exchanging meteorological data among countries of the world. All countries can make observations. Thus there is a real way for all countries to participate in a program like the WCP even though a given country might not be able to bring sophisticated instrumentation to an experiment or provide a modeling facility.

The availability of personal computers, software, and a data base composed of the country's own observational records provides the infrastructure for making excellent use of the climate data to help irrigation, design dams, and address other water resource management problems. The CLICOM will be in use in about 100 countries by the early 1990s. This project, which began as a U.S. idea, is now being funded by the United Kingdom, France, Finland, and the United States.

Policymakers and scientists working together can accomplish a great deal, as scientific issues have a way of becoming policy issues demanding solutions. The greenhouse effect, stratospheric ozone depletion, and climate change are specific examples. Policymakers and scientists alike must recognize that there always will be some scientific uncertainty involved in such issues. They must not let this uncertainty impede the decision-making process. Through the WCIP, it is anticipated that decision makers and policymakers will be able to work with scientists for the common good. That will be a very difficult task, however.

The WCP will continue for decades. Field experiments such as that concerning the tropical ocean and global atmosphere (TOGA) are long-term efforts that are scheduled to last for at least ten years. Analyses and interpretation will take another ten years, so some investigators conceivably could spend their entire research life on the project. A completion date for the WCP has not been set. The program is, in a sense, open-ended; moreover, it is stil evolving and projects are being designed and formulated even today.

The WCP is one of the building blocks for the future ICSU program known as the International Geosphere-Biosphere Program: A Study of Global Change. From the WCP will come a cadre of interdisciplinary scientists who have global horizons, an understanding of problems well beyond their own discipline, and experience in communicating internationally, independent of politics. The future is bright for the success of the WCP and the International Geosphere-Biosphere Program, but it will take many years before the harvest is gathered.

ANNALS, *AAPSS*, 495, January 1988

The Information Age in Concept and Practice at the National Library of Medicine

By HAROLD M. SCHOOLMAN and DONALD A.B. LINDBERG

ABSTRACT: The evolution of today's information age is mirrored in the growth of the National Library of Medicine's bibliographic services: from the pioneering work of the Library in the last century to develop *Index Medicus*, to developing the innovative computerized MEDLARS system in the early 1960s, to the easily searchable on-line data bases now available to health professionals. This evolution has not come about without controversy, however, as tension developed between the public sector, where the information was viewed as a social benefit, and the private sector, where it was viewed as a source of profit. The Library has fostered research and development in biomedical communications in the laboratories of its Lister Hill Center and through grants to assist in establishing the field of medical informatics. The future of biomedical communications will be profoundly affected by work now being carried out by the Library: in fields such as biotechnology, through integrative methodologies such as the Unified Medical Language System now under development, and by a grant program to improve the infrastructure for information within academic health science centers.

Harold M. Schoolman, M.D., a member of the National Library of Medicine staff since 1971, is deputy director for research and education and, at Georgetown University, holds the title of clinical professor of medicine. Previously, he was director of education services, Veterans Administration central office, and attending physician, Washington, D.C., Veterans Hospital.

Donald A.B. Lindberg, M.D., director of the National Library of Medicine since 1984, came to the Library from the University of Missouri School of Medicine, where he established one of the nation's first medical computing centers and was director of the Information Science Group for 15 years.

NOTE: Adapted in part from D.A.B. Lindberg and H. M. Schoolman, "The National Library of Medicine and Medical Informatics," *Western Journal of Medicine*, 145:786-90 (Dec. 1986).

FOR most organizations, the information age has created new opportunities and new challenges in the management of information to support their business. But for the National Library of Medicine (NLM), the management of information is its business and has been throughout its 150-year history. While much has been written in the last three decades about the information explosion, in 1881 John Shaw Billings, the then-director of the Library, wrote that the geometric progression of publication in medicine of the last few decades, if continued, would lead to the absurd conclusion that a time would soon come when it would require the services of everyone in the world not engaged in writing to catalog and index the annual output of medical literature.

For us, the information age is not characterized by either the number of people employed in the information industry or the importance of information to the country's economy and defense but rather by the emergence of a new set of tools to employ in the management of information. Nowhere is this better illustrated than in the evolution of the bibliographic services of the National Library of Medicine.

THE *INDEX MEDICUS*

Billings started the *Index Medicus* in 1879, calling it "a monthly classified record of the Current Medical Literature of the World." In his prospectus, Billings described the new *Index* as follows:

> In its pages the practitioner will find the titles of parallels for his anomalous cases, accounts of new remedies, and the latest methods in therapeutics. The teacher will observe what is being written or taught by the masters of his art in all countries. The author will be enabled to add the latest views and cases to his forthcoming work, or to discover where he has been anticipated by other writers, and the publishers of medical books and periodicals must necessarily profit by the publicity given to their productions.[1]

The first issue of *Index Medicus* listed 18,000 articles. Within fifty years its annual coverage tripled. Today, the annual coverage exceeds 250,000 articles. By the early fifties the logistics of maintaining the *Index Medicus* had become overwhelming. Currency could not be maintained and backlogs were growing rapidly.

In 1960 a new system was put into operation that handled 125,000 citations with multiple entries for each. It used tape-actuated typewriters for the repetitive printing of the unit record across the top of a Hollerith card, the keypunching of filing indicia into a reserved area of each card, sorters and collators to arrange the cards into appropriate author and subject subsets, and the use of a high-speed rotary step camera. The camera, capable of varying its aperture to fit the number of lines in the entry, would photograph the cards on film strips that could be cut into column length.

Billings would have found this a perfectly logical step, for it was he who suggested to Herman Hollerith the idea of using cards that had data represented by notches punched into the cards as a means of tabulating the 1890 census.

THE ADVENT OF COMPUTERIZATION

This early application of automation foreshadowed what might be done; it led to the specifications of the Medical

1. John S. Billings, "Prospectus," *Index Medicus*, 1(1):1 (Jan. 1879).

Literature Analysis and Retrieval System (MEDLARS). It was this system that produced the January 1964 issue of *Index Medicus.* Graphic Arts Composing Equipment (GRACE), developed specifically for the Library, typeset pages at the rate of 300 characters per second. GRACE accepted input directly from magnetic tape that had been reworded by the computer into a page-block format.

This photocopier tape contained a matrix of 226 characters etched on glass. These characters were in several sizes and fonts, with a complement of diacritical marks. Behind each character on the matrix was a high-speed flash tube; the circuitry of GRACE timed the flashing of these lights. Between the matrix plate and a nine-inch-wide roll of film there was a mirror and reciprocating lens, constantly roving back and forth, photographing one line of characters, character by character, across the entire width of the three-column page. At 1.7 seconds per sweep, it may be calculated that to compose the five volumes and 8900 pages of the 1968 *Cumulated Index Medicus,* GRACE had to labor for 150 hours.

Remarkable for its day, GRACE was a significant step in computer-controlled typesetting. By demonstrating that such a system was operationally feasible, NLM created the precursor, if not the stimulus, for computer-based publishing. This accomplishment was recognized by GRACE's retirement in 1969 to the collection of the Smithsonian Institution.

But MEDLARS was not designed solely for the publication of *Index Medicus.* From the beginning, retrieval was prominent in its specifications. By the late sixties, 16,000 searches a year were being done in batch mode on three computers that mounted the MEDLARS tapes. Medical Subject Headings had been thoroughly revised and greatly expanded. This structured, controlled vocabulary permitted in the computerized search a far greater specificity than was possible through the use of the *Index Medicus* main headings.

SEARCHING THE LITERATURE ON-LINE

The newly formed Lister Hill Center for Biomedical Communications soon began experiments with interactive on-line searching using the teletypewriter communication network (TWX) and an abbreviated file of 100 clinical journals called Abridged Index Medicus; thus the name AIM-TWX. The extension to the full *Index Medicus* file and the utilization of common communication carriers soon followed, and MEDLINE was launched. The immediate and remarkable success of MEDLINE provided a great stimulus to the common carriers to expand the coverage of their telecommunication networks rapidly. This contributed greatly to the early introduction of on-line searching of many other data bases. Today more than 4 million searches per year are made of the NLM data bases.

The on-line interactive search made the rapid identification of citations a practical reality. But citations were not the answer to a searcher's question. For that, one needed the documents to which the citations referred. To assure the availability of documents, NLM, under the aegis of the Medical Library Assistance Act of 1965, supported new Library construction, collection development, training, and special publications.

MEDICAL LIBRARY NETWORKS

Perhaps most important of all, the Medical Library Assistance Act extended to NLM the authority to establish a national communications system, including, if necessary, construction of NLM branch libraries in the states and construction or reconstruction of libraries at the medical schools. Fortunately, this authority was used wisely to build up strong local and regional libraries and to build the Regional Medical Library network that we know today.

In this system, documents that are not available at a given hospital library can be requested from larger resource libraries and the Regional Medical Library. If the material sought is not available there, NLM acts as the library of last resort. In 1985 more than 2 million documents were delivered within the network. NLM had to supply only 160,000 of these. This network routing system is being even more fully automated. A national computer-based system for interlibrary loan requests, DOCLINE, is now being implemented in stages throughout the nation.

This brief description of more than a century of the Library's involvement with medical bibliography demonstrates a remarkable, mutually beneficial, and continuing interaction between government, academia, and industry. Because the gathering and dissemination of information have always been among its prime functions, the government has always operated in an information environment. The critical change with the advent of the information age is that what heretofore had always been regarded as a social benefit now became in the eyes of many a commodity of economic importance. As that economic importance grew, so did the growth of the information industry. The inevitable result of these changes is an increasing tension between those who view information primarily as a social benefit and those who view it as a commodity. Because of the NLM's preeminent position in the application of modern computer and communications technology, it has frequently found itself deeply immersed in the growing tension.

CONFLICTING VIEWS OF HOW INFORMATION SHOULD BE TREATED

Along with the growth of awareness of information as an important economic commodity, there has grown an increasing appreciation of information as a source of power. This has given rise to a different set of tensions especially with regard to scientific and technical information, which is reflected in attempts, for various reasons, to control the flow, and especially the transborder flow, of information. Because biomedical information is almost universally viewed in a humanitarian context, however, the Library has been less affected by this set of tensions than have other institutions.

Those who view information as a commodity believe that government's role has been and should be to allow the marketplace to operate. Only when the marketplace fails should government intervene. Moreover, if intervention becomes necessary, it should be accomplished either by regulation or by subsidization. This point of view has its origin and precedent in the government's traditional relationship to agriculture and industry.

Historically, in the field of information the situation is quite different. Information is the government's business, and it operated information systems long before these issues arose. The *Index Medicus* is in its second century of continuous publication and it is far from the oldest of government data bases. The growth in the economic importance of information led to the view of some that government either should not operate information systems or, if it does, should operate them in a commercial fashion with pricing based on at least a total recovery of both generation and access costs. Those who view information as a societal good point out that Congress, in recognition of that benefit, regularly appropriates funds to generate information and to assure its availability. If the Library had to support its operations from income generated through the sale of its products and services, the availability of information would be determined by its economic viability. It would mean that the price of information would be set to maximize incomes and thus would be largely determined by those most capable of paying for it. It would mean that esoterica, whether of scholarship or rare diseases, would either be enormously expensive or, more likely, not available. It would create a situation that is fundamentally inimical to government's prime function of providing for the common good. Thus the view arose that the commodity aspect of information should be a function of value-added services, not the basic information itself.

Such tensions, early on, resulted in polarization around these two views and probably generated more heat than light. Time and subsequent events have demonstrated that they may indeed coexist to the benefit of all. The Library, for example, has joined forces with industry in a series of experiments to explore the utility of the new computer-disk/read-only-memory technology for the dissemination of biomedical information; commercial vendors provide value-added services and sell access to the Library's data bases here and overseas; and industry is more receptive to joining the government in the creation of information systems to help deal with the appropriate management of toxic substances.

MEDICAL COMMUNICATIONS TECHNOLOGY IN EDUCATION AND TRAINING

The Congress twenty years ago recognized the importance of emerging computer and communications technology in the dissemination of health science information. In 1967, by joint resolution, it created at the NLM the Lister Hill National Center for Biomedical Communications, referred to earlier.

The Lister Hill Center began to engage in experiments employing telemedicine to support health care delivery and medical education using the Advanced Technology Satellites ATS-1 and ATS-6. At the same time, the Library contracted with the Association of American Medical Colleges to assess the future role of computer and communications technology in medical education and health care delivery. The report of the Association of American Medical Colleges' committee, chaired by Dr. Eugene Stead, was published in 1971. The committee wrote with regard to computers in medicine

> that the time has come for the Lister Hill Center to exert strong leadership in the development of the computer science field as it relates to medical education and the preparation of professionals for the delivery of health care. . . . The committee believes that we will have to re-examine the problems in medical education and practice and devise solutions utilizing computer technology which will offer new approaches to education and give new patterns for professional staffing of clinical units, for the collection of clinical and laboratory data, and for clinical decision making.[2]

2. Eugene A. Stead et al., "Educational Technology for Medicine: Roles for the Lister Hill

These recommendations led to a revamping of the intramural programs of the Lister Hill Center and to the establishment of a series of research, training, and career development grants concerning the utilization of "computers in medicine."

THE MANAGEMENT OF MEDICAL INFORMATION

In 1971, the environment in the vast majority of academic health centers was barely tolerant of such endeavors. There was no locus for such activity. Indeed, there was little recognition that such a field existed. It certainly had no name. Not even the people working in the field could agree on what it should be called, let alone what it encompassed. Only years later was the name "medical informatics" adopted.

For these reasons, training grants were established that had as their primary stated objective the training of established faculty members who could return to their institutions not so much to concentrate on research careers in medical informatics—although that, of course, was not discouraged—but rather to improve the environment and to help create a greater institutional receptivity for the increasing use of automated systems in all phases of medical education and health care delivery. Research and career development grants, on the other hand, were designed to support and enhance the careers of those few existing research workers in the field.

The extent to which the training grants accomplished their avowed purpose is difficult to assess definitively. These efforts, the emergence into prominence of general information systems, microcomputers, and the rapid introduction into medicine of computer-mediated technology such as the computerized axial tomography scan, and so forth, all contributed to a radical change in environment in the decade that followed. Because of that change in environment, the training grants were changed in 1982 to research training grants with the specific purpose of training future researchers in the field.

Related to this, the NLM has recently proposed the development of centers of excellence in medical informatics. Such centers would have three major functions: coordinating medical informatics studies and skills in the overall curriculum; the conduct of research in medical informatics; and training and education in the field. Such centers of excellence will markedly enhance the recognition and career development so badly needed in the field.

COPING WITH THE CONTINUING INFORMATION EXPLOSION

It was not only the environment for medical informatics that was dramatically changing in the early 1980s. The very boundaries of clinical medicine were rapidly being pushed back to the molecular level, and with that development came an explosion of automated data bases. As the volume of medical information rapidly expanded, access to that information became more and more constrained by the system that housed it. The utility of information is to a very large extent dependent on the ease with which it can be retrieved. Two great bottlenecks to information retrieval are the difficulty of human-to-machine and machine-to-machine communication.

Center," *Journal of Medical Education*, 46(7, pt. 2):59 (July 1971).

Human-machine interface

In the late 1950s, the modern NLM classification system was rebuilt. It combined through a common set of pointers the indexing of articles and cataloging of books and monographs. This controlled, highly structured knowledge-representation scheme, Medical Subject Headings, is a very powerful tool for retrieving citations to the medical literature. Its structure and its so-called explode capability make it possible through single entries to retrieve a wide range of related material. The in-depth indexing—an average of 12 terms assigned per article—and the use of Boolean manipulators—"and," "or," and "and not"—make possible great precision of retrieval. But the system is complicated and was originally designed for relatively well-trained users.

Numerous attempts have been made by NLM and others to simplify this human-machine interface so that the system may be used by nonexperts. For certain types of searches these systems are very effective. NLM's user-friendly front end to MEDLINE, called "GRATEFUL MED," provides a transparent, menu-driven use of Boolean operators "and" and "or" with free language entries that are searched both as free text and as vocabulary controlled by Medical Subject Headings.[3] Even in its simplest form, utilizing only selected features of the power of the MEDLARS system, it will retrieve enough relevant citations to satisfy many queries. On the other hand, the more experienced user can operate through another mode of GRATEFUL MED to bypass the menus and have all the power of MEDLARS command language searching restored.[4]

Machine-machine interface

An obvious improvement in user-friendly front ends is to have them able to communicate with more than one data base. The Chemical Substances Information Network (MICRO-CSIN) is such a system. It knows the protocols and can therefore automatically log into any of the hundreds of data bases available from nine main vendors. Because its basic purpose, as the name implies, is to search chemical systems, searching across data bases is reasonably effective. This machine-to-machine communication is facilitated by the presence of unique identifiers in chemistry such as the chemical name and the Chemical Abstracts Service registry number. These unique identifiers make it possible to find the chemical of concern regardless of the data base and the data-base structure. A small number of key concepts, such as toxicity, physical and chemical properties, and the like, then make it possible to retrieve the record of interest.

CREATION OF A UNIFIED MEDICAL LANGUAGE SYSTEM

Unfortunately, for most of biomedicine there are no unique identifiers. Even where data bases have a shared subject domain, they employ different knowledge representation schemes using different indexing philosophies and different vocabularies. In these instances, the transfer of a concept from one data

3. Elias Abrutyn, "Software: Grateful Med," *Annals of Internal Medicine*, 105(2):321 (Aug. 1986).

4. GRATEFUL MED is available for $29.95 from the National Technical Information Service, U.S. Department of Commerce, Springfield, VA 22161; its order number is PB86-158482.

base to another is extremely uncertain, being subject to all the vagaries of normal discourse without a human to interpret them. There are three possible solutions to this problem: adopt a standard vocabulary, interpose an expert system that simulates the human intermediary, and create some canonical knowledge-representation scheme into and from which all others can be transposed.

The NLM believes the first solution to be unacceptable to the community. A combination of the second and third is being studied with the help of the academic community and professional societies. This combination has been called a unified medical language system. It is viewed as a long-term—and costly—project from which many intermediary benefits may be derived.

ORGANIZATIONAL REQUISITES FOR THE MANAGEMENT OF HEALTH CARE INFORMATION

Solving the technical problems of knowledge representation, unified medical language systems, and machine-to-machine communication is not all that is required. Institutional restructuring to manage effectively the complex information activities within health centers must also occur. For that reason, NLM instituted a program in 1983 to support the study, planning, and effectuation of Integrated Academic Information Management Systems.[5] This program was initiated with planning grants and contracts to four academic health centers; four additional awards were made a year later. Experimentation is now going on and implementation of several models will occur in the next few years.

The Integrated Academic Information Management Systems program was started because the Library recognizes that information management is a pervasive issue among the health science centers. Research results, clinical records, student records, laboratory results, library holdings, national data bases, and institutional material are only a few of the information sources that frequently need to be unified in a comprehensive manner. Thus the first objective of the program was to raise the consciousness of health professionals to the need for the integration of information sources and thus for the rationalization of the management of those resources. From this should come institutional policies that establish who shall be responsible for what data, who shall have access to what data, how the data management will be financed, what hardware and software compatibilities are essential, and how the important issues of privacy and security will be managed. Finally, the program should stimulate the creation of several operational models that may be used as laboratories to study these complex institutional information-management problems.

We have, in this brief recapitulation of the history of an information organization in the emerging information age, attempted to indicate how the rapidly changing computer and communications technology has not only affected the organization's internal operations but also its relationship with academia and industry. The NLM supports research in information management—medical informatics—to try to make as rapid as possible the applications at the bedside of the results from the research laboratories.

5. Nina W. Matheson and John A.D. Cooper, "Academic Information in the Academic Health Sciences Center: Roles for the Library in Information Management," *Journal of Medical Education*, 57(10, pt. 2) (Oct. 1982).

MEDICAL INFORMATICS IN THE ERA OF THE NEW BIOLOGY

Now, at a rapidly increasing rate, the view of biology is shifting to the molecular structure of genes and their protein products. Research in the life sciences is becoming increasingly dependent upon tools to store and manipulate large amounts of data on the behavior and structure of macro-molecules. The ability to measure and change events occurring on a molecular level is particularly significant where the development of techniques to sequence, clone, and remodel genetic material is leading to the control of life processes with a precision never before known. The continued pursuit of this knowledge is as much an information-processing problem as it is a problem in biology. Unfortunately, the institutional infrastructure for the support of the information-processing part of these endeavors is far less defined and developed than that for the biology. NLM already plays a crucial role in its bibliographic control of the published literature. The management of the rapidly emerging banks of data need now to be organized and integrated.

The complexity and size of biotechnological information stagger the imagination and make it different from other scientific information. Biotechnological information is growing at an enormous rate, and there is certainly a lot of ground to cover. Within each of us tens of thousands of individual genes control special life processes. Three billion units of DNA make up the human genome and only 0.01 percent have been sequenced. The nucleic acid sequences in the gene constitute a language that governs the building of proteins that regulate the metabolic functions of the body. Those proteins are made up of amino acids that are also arranged in a specific sequence. Data bases of these sequences have been started, but these data bases are swamped. The best known, GenBank, was set up under contract with the National Institutes of Health and co-funded by a number of federal agencies. By mid-1986 only 54 percent of the data published in 1985 had made its way into the data bases. This is because the rate of publication has grown so rapidly—from 1 sequence composed of 76 bases in 1965 to a total of 11,552 sequences composed of nearly 10 million bases by 1986. What is at stake is not merely the convenience of researchers searching for scientific papers but rather progress in biotechnology around the world—our understanding of life at the molecular level and hence our knowledge of health and disease.

The numerous data bases that do exist use different information systems and different computer languages and have different structures. The limitations are not only the logistics of entering the enormous amount of data into automated systems. Even more difficult is retrieving the information when it exists in multiple places. The whole biotechnology information system is so overloaded that advancement in the field is progressively impeded. The future development of biotechnology is at this moment more of a problem in information access and management than of biology. With this problem in mind, Representative Claude Pepper introduced a bill to create a National Biotechnology Information Center at the NLM.[6]

Working with the laboratories from which information comes, experts at such a center would create new computer

6. The bill was introduced in the 99th Congress as H.R. 5071 and reintroduced in the 100th Congress as H.R. 393.

information systems that would have consistent terminology allowing researchers to obtain answers to questions and to share research results quickly. In these information systems, research data would be stored in such a way that information from one source could be linked to other related findings and to other data bases. Investigators could ask for information from one computer system and that computer system would automatically search for an answer not only in its own knowledge base but in other research data bases as well.

With such important issues facing it, the Library is now engaged in a careful look to the future. NLM has just completed the initial phase of a long-range planning activity in which it has had the views and advice of five panels of distinguished outside experts. The topic of research in medical informatics was carefully and specifically considered by one of these panels.

The final recommendations are now under review by the Board of Regents. It is already clear, however, that the board readily accepts the importance of information management in biotechnology, that automated information systems and knowledge bases are critical elements in its vision of the future, and that research in medical informatics is essential in assuring the best of these future developments.

ANNALS, *AAPSS*, **495**, January 1988

On-Line Research-Support Systems

By MURRAY ABORN and ALVIN I. THALER

ABSTRACT: This article provides an overview of trends in electronic communication among research scientists, with particular attention to the Experimental Research in Electronic Submission (EXPRES) initiative at the National Science Foundation (NSF). NSF is an independent agency of the federal government authorized to initiate and support basic scientific research and to strengthen the nation's scientific research potential and education in the sciences. EXPRES is a conceptual and technological project attempting to facilitate the exchange of compound scientific documents between research sites possessing different kinds of computer hardware and software. In its most fundamental aspect, EXPRES is about scientific collaboration as well as communication. This dual process is exemplified in the development of basic research proposals by scientists at universities, and the subsequent procedures by which such proposals are submitted, evaluated for scientific merit by the scientists' peers, and ultimately funded or declined. EXPRES holds the promise of not just streamlining but also enhancing the support systems by which today's scientific research is carried out.

Murray Aborn is senior scientist in the Division of Social and Economic Science, National Science Foundation. He also serves as program director of that division's Measurement Methods and Data Improvement Program. He received his Ph.D. from Columbia University.

Alvin I. Thaler, recently of the National Science Foundation's Mathematical Sciences division, is currently program manager for Experimental Research in Electronic Submission, and program director for Instrumentation in the foundation's Directorate for Computer and Information Science and Engineering. He received his Ph.D. from the Johns Hopkins University.

WE are much indebted to Derek Price for attempts to apply the methods of scientific investigation to the study of science itself. Over a period of many years, he has contributed greatly to the measurement of scientific publications as a gauge for the growth of science qua science and to the role played by science in the political economy of societies, both past and present. Although he did not coin the term "big science," he explicated its relationship to and emergence from the realm of little science, and he adumbrated the sociological and political implications of this shift for the information age in which science functions today.[1]

THE GROWTH OF BIG SCIENCE

A quarter-century ago, Price pointed out that

> the science we have now so vastly exceeds all that has gone before, we have obviously entered a new age that has been swept clear of all but the basic traditions of the old. Not only are the manifestations of modern scientific hardware so monumental that they have been usefully compared with the pyramids of Egypt and the great cathedrals of medieval Europe, but the national expenditure of manpower and money on it have suddenly made science a major segment of our national economy.[2]

Eight years earlier, the National Science Foundation (NSF), newly created along then-conventional notions of pure science as essentially small scale and individualistic in character, was already finding itself becoming deeply enmeshed in bigger science, a style of operation destined to command an increasing portion of its attention and budget. The emergence of new areas of study such as oceanography, geophysics, and the atmospheric sciences tended to demolish traditional disciplinary boundaries. It not only required adjustments in NSF's original patterns of evaluation and support but demanded expensive equipment and facilities plus the cooperation of widely dispersed investigators—all prominent features of science writ large.[3]

The pull of technology

Because big science is often characterized by large facilities and expensive pieces of equipment—"big ticket items," in the parlance of federal grant administration—it can increase the pull that technology normally exerts on the development of science. In a volume celebrating the inauguration of NSF's *Science Indicators* series—a series of biennial reports designed to provide policymakers with a broad base of data about the state of U.S. science and technology—Derek Price noted that in the historical development of science, "the primary strong flow of information runs from the technology to the scientists; only secondarily, and after understanding has been achieved, does this new wisdom succeed in producing change and innovation in the technological process."[4] Price went on to say,

> Probably because far too high a proportion of the historians—and indeed, the sociologists of science—have been theoreticians rather than experimentalists, there exists a chronic underestimation of the part played

1. Derek de Solla Price, *Little Science, Big Science* (New York: Columbia University Press, 1963).

2. Ibid., p. 2.

3. J[ames] Merton England, *A Patron for Pure Science* (Washington, DC: National Science Foundation, 1982), pp. 279-310.

4. Derek de Solla Price, "Towards a Model for Science Indicators," in *Towards a Metric of Science: The Advent of Science Indicators*, ed. Yehuda Elkana et al. (New York: John Wiley, 1978), pp. 69-95.

by the craft of experimental science. It is precisely in the region populated by instruments, big and small machines, gimmicks, techniques and goos, that the interaction runs high, hectic, and startling, and that useful technological innovation is endemic.[5]

Nothing said before or afterward so beautifully encapsulates so much of the motivation behind research-support programs attempting to utilize technological experimentation as a means of creating new scientific paradigms. One such program, the Experimental Research in Electronic Submission (EXPRES) initiative at NSF, being carried out in collaboration with participating universities, is described and discussed in the section that follows.

BIG SCIENCE AND THE FEDERAL SUPPORT OF RESEARCH

In their now-classic work on the coming emergence of a "network nation," which is to say, a nation in which electronic modes of communication become the ordinary forms of social and intellectual exchange among all elements of the citizenry, Hiltz and Turoff gave special attention to the effects on science and technology of what they broadly dubbed "computerized conferencing." In computerized conferencing, distance is eliminated as a factor in communication. Networks capable of vastly increasing both the speed and the quantity of information transmission were envisioned as decreasing the isolation of scientists in geographically remote institutions and increasing the connections between scholars sharing the same specialized interests.[6]

Hiltz and Turoff saw computerized conferencing as particularly well suited to the needs of "invisible colleges," groups of researchers linked into unofficial collegia of common interest, utilizing rump conferences, working papers, and other modes of informal correspondence as their media of communication. Fifteen years earlier Derek Price identified such groups and described their emergence as indicative of the transition from little science to big science.[7] Now Hiltz and Turoff predicted that the use of computerized teleconferencing not only would produce greater togetherness but would change the social structure of these groups and could affect the very development of different scientific specialties.

One singularly important accompaniment of growth in the scale of science is growth in the costliness of research, and one overwhelmingly important consequence of higher cost is the reliance this creates on governmental sources of support. Since World War II, federal funding agencies have been the guarantors of advancement in many fields of science, and in the case of agencies like NSF or the National Institutes of Health, research grants have been the mainstay of progress in fundamental knowledge across the spectrum of scientific disciplines.

Envisaging electronic submission and review procedures

In a remarkable prognostication of things to come, Hiltz and Turoff described the future prospects for computerized conferencing in the grant-review process.[8] As they pointed out, scientific research was becoming so specialized

5. Ibid., p. 71.
6. Starr Roxanne Hiltz and Murray Turoff, *The Network Nation: Human Communication via Computer* (Reading, MA: Addison-Wesley, 1978).
7. Price, *Little Science, Big Science*, pp. 62-91.
8. Hiltz and Turoff, *Network Nation*, pp. 214-50.

and so dynamic that scholars from the same invisible college were often the only ones capable of evaluating the merit of research in that specialty. Though seemingly at odds with common-sense notions of objectivity, peer review had for some time established itself as the most successful method of maintaining quality control in science, and this method was firmly entrenched in agencies vested with responsibility for the support of scientific research.

In addition to providing a mechanism for quality control, peer review performs other important services that accrue to the benefit of scientific research and the expenditure of public moneys. Among them are (1) the motivation to prepare well-thought-through research plans that uphold the author's prestige in the eyes of his or her colleagues, (2) the furnishing of guidance to federal granters with respect to research that simply duplicates what has already been done, (3) the furnishing of constructive criticism for improving the quality of research that would otherwise remain substandard, and (4) the creation of channels of communication that keep researchers aware of current developments in their special areas of investigation.[9]

Hiltz and Turoff proposed that putting peer review procedures on line—that is, conducting the entire process by computerized teleconference techniques—would greatly increase the benefits to applicants, agencies, and reviewers as well as overcome certain built-in limitations to conventional procedures. For instance, in place of having applications for research support evaluated by experts acting independently of each other and corresponding by inflexible, ordinary mail, proposed research plans would be scrutinized by experts interacting in real time and behaving in the manner of a panel. In the teleconferencing style of review, overlapping competencies can be brought into play, controversial points of view discussed and resolved, unclarities raised and clarified by direct contact with the applicant-researcher, and rank orderings of projects according to criteria of merit accomplished at the time when comparisons are freshest and clearest in the minds of all those concerned with the best apportionment of funds.[10]

The emergence of EXPRES

A number of federal agencies have looked into the possibilities for implementing systems of the kind envisaged by Hiltz and Turoff as an alternative to the conventional methods of proposal submission and evaluation. NSF was among the first to move decisively in this direction. In June of 1986, NSF announced that it was seeking competitive proposals for projects involving the conduct of experimental research in electronic submission. This research will certainly lead to and include experimentation aimed at expanding existing mechanisms for collaboration and communication between scientists and scientific groups. The thrust of the announcement was heavily technological, placing the immediate emphasis on new technologies permitting the exchange of compound documents between dissimilar hardware and software environments. Compound documents are the rule rather than the exception in most areas of scientific communication, where text may appear in mixed-mode presentation, with numerical and symbolic data, images of all sorts, graphs, photographs, and, looking to the future, animation and voice. Also

9. Ibid., p. 238.

10. Ibid., p. 239.

residing in this program, however, in addition to the technological emphasis, are the seeds of change in work styles, knowledge exchange, and social structure that Hiltz and Turoff hinted at in their magnum opus.

The advent of EXPRES may have been presaged a year earlier in the congressional testimony of John Seely Brown, then director of the Intelligent Systems Laboratory, Xerox Palo Alto Research Center.[11] In written testimony submitted to the Task Force on Science Policy of the House Committee on Science and Technology, Brown spoke of the new communication technologies in terms of enhancing our ability to tap the creative and analytic potential of scientific communities.

"One of the crucial problems of increasing specialization in the sciences," he wrote, "is the need for effective communication between individuals and subgroups and for preserving group knowledge over time and across projects." He went on to say:

> For example, even within relatively small, cohesive research groups, the need looms large for collaborative problem-solving, collaborative writing, and group brainstorming. Appropriately designed tools to enhance collaboration through the sharing of files, the structuring and archiving of group interactions, online editing and annotation of documents, and so on, are sorely needed.

"Seen as a research problem," he concluded,

> technological support of collaboration is just the sort of area that requires interdisciplinary scientific work: not only the technical but the sociological aspects of collaboration require a deeper understanding than we currently have. However, I believe that research in this area can yield impressive results in terms of increased intellectual productivity.[12]

The scope of EXPRES

The EXPRES initiative is expected to have substantial influence on the manner in which scientists prepare their documents and communicate with their colleagues and peers. While the impetus for the initiative arose from NSF's desire to increase the efficiency of its proposal-processing activities, the current view is that present-day proposal submission, processing, and evaluation procedure provides an excellent vehicle through which one can tackle the larger issues of enhanced communication and collaboration.

One of the aims of the activity is to catch the attention of the nation's scientific and engineering research communities and to make them aware of the capabilities and potentials of this emerging communication and collaboration technology. Therefore, as part of the EXPRES initiative, the two major participating universities—the University of Michigan and Carnegie-Mellon University—have begun publication of a newsletter "to share with the broader scientific community both the visions and the accomplishments of the EXPRES project."[13]

To quote from the first issue of the newly published *EXPRES Newsletter*:

> While computing technology has influenced many aspects of scientific research, there are still many new avenues to explore. Communication, collaboration, and information acquisition and dissemination still rely on technolo-

11. John Seely Brown, "The Impact of the Information Age on the Conduct and Communication of Science," Testimony before the Task Force on Science Policy, Committee on Science and Technology, House of Representatives, U.S. Cong., 1st sess., 12 Sept. 1985.

12. Ibid., pp. 7-8.

13. *EXPRES Newsletter*, 1(1):7 (June 1987).

gies such as paper and printing, video and audio tape, the mail, the telephone, and face-to-face interaction. Many scientists have discovered text-only electronic mail, and special software has been developed to make possible either synchronous or asynchronous text-only computer conferencing. While electronic mail has had a significant impact on the communication patterns of certain groups of scientists, traditional technologies still dominate.

The newsletter then goes on to explain the purpose of EXPRES in terms of the following questions: "To what extent can present and near-future technology have a major impact on the communication, collaboration, and information acquisition and dissemination habits of scientists and engineers? What problems must be overcome to make this technology a viable alternative?"[14]

The streamlining potential of EXPRES

Apart from its larger but necessarily more distant social, structural, and intellectual goals, EXPRES is seen as a cost-effective means of streamlining and otherwise improving current procedures by which research plans are conceived, fashioned, transmitted to funding agencies, evaluated, and supported or returned for reconsideration or abandonment. Preparing a grant application or responding to a proposal solicitation is a highly detailed and complex business. Simply maintaining mastery over the great volume of relevant programmatic materials issued regularly by funding agencies can be confusing and intimidating. In addition, it is usually important to keep abreast of ancillary reports and related forms of support documentation frequently commissioned from scientific advisory bodies.

Because of this, professional associations of many kinds publish guides, manuals, and other reference works to aid the scientist seeking support for his or her research. For example, one such guide contains a primer for scholars approaching the federal research-support system for the first time. It begins by warning that those familiar with the system speak in idiosyncratic tongues filled with abbreviations, acronyms, and references to incredibly large sums of money.[15] Getting down to more serious business, however, the primer instructs the reader about the importance of understanding the "mission" or fundamental purpose of the agency to which a research project is to be submitted, the importance of regular contact and communication with agency staff, some basic pointers about the kinds of methodological detail usually required, the general outlines of proposal processing in the mechanical sense, certain standards demanded by peer review, and a breakdown of the various categories of support offered by most funding agencies.

Although unquestionably instructive, publications of this type can do no more than scratch the surface of the complicated, diverse, and dynamic procedures by which today's complex of funding sources sustain the scientific research enterprise. The advanced communication and management tools inherent in EXPRES, on the other hand, hold the promise of simplifying and facilitating the entire research-support system, making it more responsive to researcher

14. Ibid., p. 1.

15. Robert P. Lowman and E. Ralph Dusek, "Federal Research Support: A Primer for Proposal Writers," in *Guide to Research Support*, ed. E. Ralph Dusek et al. (Washington, DC: American Psychological Association, Scientific Affairs Office, 1984), pp. 41-53.

needs and agency accountability requirements and more cost-effective all told.

THE TECHNOLOGY BASE

Reaching the goals of EXPRES will require widespread availability of somewhat more functional personal computers, commonly called "workstations," along with easy-to-learn, user-friendly software that offers practical assistance to the difficult tasks involved in the conduct of sound scientific research. Major developments in the marketplace are making it increasingly likely that such high-performance workstations as are called for by EXPRES will be within the financial means of scientists at many academic institutions, while software systems to use them appear to be on the horizon, though currently at an early stage of development.

It is expected that many more of these inexpensive high-performance workstations will become available within the next year or so and that this next generation of personal computers will furnish the most advanced document and graphics design tools, provide intelligent tutorial systems for their use, and combine these capabilities with immense computational power and enormous memories. Large display screens will permit the scientists simultaneously to view multiple windows containing pages of text, maps, flowcharts, or symbolic equations; and, because the screen will be made up of an array of 1 million dots, called "pixels" (picture elements), animated depictions of laboratory experiments or architectural designs will be possible.[16] Perhaps more significant, users of such workstations will be able to have their computers perform several tasks simultaneously.

16. John P. Crecine, "The Next Generation of Personal Computers," *Science* (28 Feb. 1986), pp. 935-42.

EXPRES at participating universities

Carnegie-Mellon University is one of two institutions currently collaborating in the development of EXPRES and, in doing so, putting its existing campus computing system at the service of the project's technological base. The existing system is called "Andrew" and is designed to support multimedia document preparation, electronic mail and conferencing, and instructional software development. Andrew includes a distributed file system that is accessible from any workstation; users can access all files from any workstation on campus that is connected to the network. The Carnegie-Mellon system will be able to display combined text, diagrams, and mathematical equations as well as customized, user-defined objects. Extensions and enhancements of Andrew are planned that will permit the inclusion of spreadsheets, annotations, footnotes, cross-references, indexes, and bibliographies.

The University of Michigan's role in EXPRES builds upon a multimedia messaging system called "Diamond," which allows a user to compose an electronic message made up of structured text, images, graphics, spreadsheets, and voice. As part of the EXPRES project, plans have been laid to incorporate a technical composition system called TeX, which can format pages that have complex layout requirements such as footnotes, references, tables, and indexes. TeX will also allow the inclusion of mathematical symbols and formatted equations.

In addition to Carnegie-Mellon University and the University of Michigan, some 17 other universities from around the nation have agreed to become involved in EXPRES by helping to test the system as it develops.[17]

PROBLEMS AND PROSPECTS

Increasingly, scientific explorations are being carried out by teams of researchers rather than single investigators, and more and more of those teams are composed of scientists from geographically dispersed sites. The ability to confer over long distances via any conceivable type of media, in real time as well as asynchronously, has thus created considerable excitement among many in the scientific community. This enthusiasm, however, must be tempered by caution over the possibility of doing harm by prematurely degrading the currently working system or by overlooking the problems that a new system might pose to the security, means of authentication, guarantees of privacy, or the integrity and balance of the review and evaluation process we have now.[18]

Given these concerns, it may be appropriate to conclude by taking a look at two diverse instances of group behavior where computer-based communications were involved. The first is a real-world case history in which a computer model of deep ocean mining was used to help resolve some of the complex issues entailed in negotiating an internationally acceptable law of the sea.[19] Here parties representing diverse—and normally divisive—national interests were willing to join together in examining complicated technical, legal, and cultural issues on the basis of iterative consultations with a computer model, a model capable of projecting the consequences of alternative legal and political formulations. An evaluation of the experience showed hopeful signs that this novel form of negotiation, utilizing advanced communications technology, can be of genuine assistance in resolving very difficult social conflicts.

The second instance is a more highly structured one, involving the conduct of controlled experiments in electronic messaging to determine the impact of different communication systems on group decision making.[20] Here the outcomes are not so sanguine. They suggest that different system designs produce very different behavioral effects, that technologies can worsen as well as improve decision making, that systems must be tailored to particular communication needs and intents, and that continued research on effects is crucial to the process.

17. *EXPRES Newsletter*, 1(1):4-6 (June 1987).

18. Ibid., p. 4.

19. Donald B. Straus et al., "Computer-Assisted Negotiations: A Case History from the Law of the Sea Negotiations and Speculation Regarding Future Uses," in *Computer Culture: The Scientific, Intellectual, and Social Impact of the Computer*, ed. Heinz R. Pagels, Annals of the New York Academy of Sciences Series, vol. 426 (New York: New York Academy of Sciences, 1984), pp. 234-65.

20. Sharon Murrel, "Computer Communication Systems Design Affects Group Performance," in *The New World of the Information Society*, ed. John M. Bennett and Trevor Pearcey (New York: Elsevier Science, 1985), pp. 279-85.

ANNALS, *AAPSS*, **495**, January 1988

Machine Cognition and the Downloading of Scientific Intellect

By MURRAY ABORN

ABSTRACT: This article concerns itself with what artificial intelligence (AI) may have in store for how science will be conducted and communicated in the not too distant future. It highlights some of the AI literature to show that computers have been successfully programmed to perform many of the procedures involved in scientific decision making and problem solving. It briefly reviews the ongoing debate over whether systems designed to exhibit intelligent behavior really think in the same sense that humans do. Finally, it reports on the possibility that more of the process of science will in the future be carried out by such systems as a result of research linking AI with progress in cognitive science.

Murray Aborn is senior scientist in the Division of Social and Economic Science, National Science Foundation. He also serves as program director of that division's Measurement Methods and Data Improvement Program. He holds the Ph.D. from Columbia University, is a member of Sigma Xi, and is a fellow of the American Psychological Association, the American Statistical Association, and the New York Academy of Sciences.

NO publication on the information age would be complete without some discussion of artificial intelligence (AI). AI is the intellectual progeny of the information age and seems a good candidate to become its successor. Indeed, the succeeding era could one day be named the age of the artificial, with deference to H. A. Simon's *Sciences of the Artificial*, a seminal work thematically dedicated to the proposition that the methods and findings of the natural sciences may be extended to artificial objects.[1]

HISTORICAL HIGHLIGHTS

First speaker: "Speak up!"
Second speaker: "Men are all alike."
First speaker: "In what ways?"
Second speaker: "They're always bugging us about something or other."
First speaker: "Can you think of a specific example?"
Second speaker: "Well, my boyfriend made me come here."
First speaker: "Your boyfriend made you come here?"
Second speaker: "He says I'm depressed much of the time."
First speaker: "I am sorry to hear that you are depressed."
Second speaker: "I need some help, that much seems certain."
First speaker: "What would it mean to you if you got some help?"
Second speaker: "Perhaps I could learn to get along with my mother."
First speaker: "Who else in your family takes care of you?"
Second speaker: "My father."
First speaker: "Your father?"
Second speaker: "You are like my father in some ways."
First speaker: "What else comes to mind when you think of your father?"
Second speaker: "Bullies."
First speaker: "Does that have anything to do with the fact that your boyfriend made you come here?"

Conversation

The preceding specimen discourse is, as the reader will surely have guessed, taken from a dialogue between a patient and her psychotherapist. It is reminiscent of a therapeutic technique common in the 1960s. There is nothing remarkable about the exchange except that one of the parties to it is a computer program.[2] The program, called "ELIZA," was written to mimic the psychotherapeutic process. It did so surprisingly successfully, largely owing to its choice of method—the noninterventionist, Rogerian approach in which the therapist behaves much in the manner of a sounding board.

ELIZA is precious in the annals of AI because it is an excellent example of early attempts to turn the computer into a natural language processor and apply this capability toward practical ends. But it has also received attention for the recoil that its creator exhibited when he discovered the naive subservience with which the device was being espoused, even by elements of the scientific community. Ten years after he wrote ELIZA, Joseph Weizenbaum was to write, "Whatever intelligence a computer can muster, however it may be acquired, it must always and necessarily be alien to any

1. Herbert A. Simon, *The Science of the Artificial* (Cambridge: MIT Press, 1969).

2. Joseph Weizenbaum, "ELIZA—A Computer Program for the Study of Natural Language Communication between Man and Machine," *Communications of the Association for Computing Machinery*, 9:36-45 (1966).

and all authentic human concerns. The very asking of the question, 'What does a psychiatrist know that we cannot tell a computer?' is a monstrous obscenity . . . a sign of the madness of our times."[3]

Comprehension

Weizenbaum's reaction was not based solely on moral rage. Early natural language processing programs were written under the assumption that useful question-answering procedures could be created employing nothing more than the syntactic information in the sentence together with the denotative meaning of a restricted set of words. But not everyone was deluded in this regard. It was clear to some researchers early on that sans cognition—in the absence of extensive world-knowledge—natural language processors could never be made into anything like fluent conversationalists, to say nothing of performing tasks requiring translation or interpretation.

Thus scientific research in AI increasingly turned to natural language processing systems that took semantic considerations into account, and that trend produced programs such as LUNAR, which has been used to answer questions about rocks brought back from the moon; SHRDLU, which is able to simulate, albeit in a very primitive fashion, a robot manipulating objects on a tabletop; MARGIE, which, if fed certain English sentences, can make inferences about the sorts of actions likely to eventuate; and SAM and PAM, which can accept as input text containing complete—albeit exceedingly simple—narratives and either paraphrase them or draw meaningful inferences from them.

Consultation

So-called expert systems are an AI success story. They are essentially consultative in character, serving as decision-making aides in which the computer is endowed with the same sort of expertise that a practitioner or researcher might obtain from a person who is highly skilled in some domain of knowledge. One of the first such systems, MYCIN, is a good example. MYCIN is designed to provide an attending physician with advice on diagnosis and therapy for infectious diseases that might arise suddenly during hospitalization and require immediate action. Evaluations of MYCIN have given results that tend to satisfy a criterion many researchers would accept as authoritative—the so-called Turing test—in that MYCIN compares favorably with the performance of human experts. This comparison has encouraged the development of extensions of the system, such as NEOMYCIN, which are aimed at providing tools for medical education and training.[4]

Common sense

Although expert systems are now considered to be essential components of advanced decision-making systems in many areas of science, medicine, and engineering, their development is hampered by serious limitations even apart from those having to do with the tech-

3. Joseph Weizenbaum, *Computer Power and Human Reason: From Judgment to Calculation* (San Francisco: Freeman, 1976). The quotation cited here is taken from Grant Fjermedal, *The Tomorrow Makers* (New York: Macmillan, 1986), pp. 106-7.

4. Avron Barr and Edward A. Feigenbaum, *The Handbook of Artificial Intelligence* (Stanford, CA: HeurisTech Press, 1982), 2:184-92, 267-78.

nology of downloading skill-level human expertise.

Expert systems are meant to operate in the real world rather than in highly controlled and restricted laboratory environments. They must therefore be capable of dealing with real-world social and psychological problems, be able to accumulate and integrate rapidly introduced new information, and both possess and know when to apply common sense. John McCarthy of Stanford University, one of the founders of AI, gives the last of these requirements top priority. Speaking at a 1984 conference and using MYCIN as a case in point, McCarthy repeated what he had been saying since 1958, namely, that acquiring a better understanding of commonsense facts and methods is the key problem facing artificial intelligence.[5]

Simulation

The limitations of expert systems notwithstanding, these systems are considered to be among the building blocks now available for very large-scale applications of AI. An excellent illustration is a recent article in *Science* that assigns AI a central role in the management of natural resources.[6] Although appropriate warnings are sounded concerning their limitations and the consequent dangers of receiving bad advice, existing agricultural, entomological, geographic, and zoological expert systems are envisioned as functioning in an integrated fashion alongside other AI tools and techniques to simulate the behavior of entire ecosystems. Natural language processing and some of the other methodologies of AI discussed earlier in the present article are described as integral components of the proposed resource management scheme. The scheme also incorporates techniques and products of AI such as robotics and image processing, very briefly touched upon later in the present article.

Problem solving

Heuristics, or the use of indeterministic search strategies, seeks to provide shortcuts that avoid the necessity for heavy computation in automated problem solving. This line of research is prominently associated with Allen Newell and Herbert Simon of Carnegie-Mellon University. It began at the virtual outset of the field and stands today at its forefront. It is a methodology that has already expressed itself in important practical developments—expert systems, for example—and that is sure to affect science even more profoundly in the future.

Newell and Simon's General Problem Solver appeared in 1957 as a first attempt to mechanize problem solving in a general, rather than a domain-specific, mode. It became the subject of ten years of work aimed at improving the program's versatility in playing games, proving theorems, and solving puzzles.[7] Its major impact, however, was goal directional, not technical. Some of the conclusions reached by Newell and Simon

5. John McCarthy, "Some Expert Systems Need Common Sense," in *Computer Culture: The Scientific, Intellectual, and Social Impact of the Computer*, ed. Heinz R. Pagels, Annals of the New York Academy of Sciences Series, vol. 426 (New York: New York Academy of Sciences, 1984), pp. 129-37. For a synopsis of authoritative viewpoints on the subject of implanting commonsense knowledge, see the Panel Discussion in ibid., pp. 138-60.

6. Robert N. Coulson, L. Joseph Folse, and Douglas K. Loh, "Artificial Intelligence and Natural Resource Management," *Science*, 237:262-67 (July 1987).

7. Barr and Feigenbaum, *Handbook of Artificial Intelligence*, 1:113-18.

were influential in turning many researchers away from the production of empty-headed applied systems or simply trying to make computers smart. It also helped to establish an ideology that places the emphasis on AI's being a laboratory for studying the fundamental nature of human intelligence.

Rediscovery

No matter how one views where the emphasis really lies, AI already possesses powers that rival what have always been considered to be among the higher-order human cognitive abilities. Almost as significant, the speed with which it has acquired these powers is no less than breathtaking. The span of time we are dealing with in surveying the development of AI—albeit ever so cursorily—is only about thirty years. From the General Problem Solver in 1957 to a program aptly named BACON in 1979-80, the field has stepped—or, more appropriately, leaped—into science's most revered domain. Here is how the creator of BACON, Herbert Simon, describes the achievement:

> The process of scientific discovery is sometimes driven by data, sometimes by theory, and perhaps most often by a combination of both. BACON explores the data-driven aspects of discovery. Given data on the distances of the planets and the periods of their orbits, BACON quickly rediscovers Kepler's third law. Given data on an electric circuit with different lengths of resistance wire, it rediscovers Ohm's law. In a similar manner, it rediscovers the inverse square law of gravitational attraction, Snell's law of diffraction, Black's law of temperature, and many of the chemical laws discovered by Dalton, Gay-Lussac, Avogadro, and Cannizzarro in the first half of the nineteenth century.[8]

Two additional quotations are interjected at this juncture because they so nicely bridge the preceding historical highlights to the next section of this article, which, for better or worse, is headed "Thinking Machines." The first quotation is taken again from Herbert Simon. In it, Professor Simon reveals the scientific significance of BACON: "From a psychological standpoint, there cannot be any differences between processes of original discovery and processes of independent rediscovery."[9]

The second quotation is more recent. It, too, contains a reference to the scientific significance of BACON's ability to rediscover physical laws, to wit: "We can conclude that we don't have to postulate any kind of mysterious processes, any kind of fundamentally unknown—much less unknowable—processes to account for what in humans we like to call creativity. Hence, we can't establish the creative process as an Iron

8. Herbert A. Simon, "Studying Human Intelligence by Creating Artificial Intelligence," *American Scientist*, 69:300-309 (May-June 1981). BACON is no longer alone in this area of effort. For instance, *Research News*, 37(5-6):5 (May-June 1986), reports that Manfred Kochen at the University of Michigan is developing PLAUSIBLE MATHEMACHINE, a program that will be able to learn geometric principles on its own. It will attempt to rediscover the Pythagorean theorem, for example.

9. Simon, "Studying Human Intelligence," p. 307. There are some AI programs in the nature of expert systems that may be said to perform original discovery rather than rediscovery. One called TETRAD supplies assistance in the area of statistical modeling. TETRAD automatically filters through large quantities of nonexperimental data in search of causal explanations, helping the researcher to select among the great number of alternative causal models that nonexperimental research can generate. A detailed description of TETRAD is contained in Clark Glymour et al., *Discovering Causal Structure* (Pittsburgh, PA: Carnegie-Mellon University, Laboratory for Computational Linguistics, 1986).

Curtain that limits what computers can do."[10]

THINKING MACHINES

Even if we were to accept as fact that there is no apparent limit to what computers can do, is this the same as saying that computers will one day possess minds indistinguishable from our own? At a 1984 conference sponsored by the New York Academy of Sciences, Marvin Minsky of the Massachusetts Institute of Technology put it this way: "If you made a machine that looks as if it thinks, would it really think?"[11] On that same occasion, John McCarthy provided one kind of answer: "A program that simulates thirst is not going to be thirsty. For example, there is no way to relieve it with real water."[12]

The idea of electronic computers as thinking machines goes back to a book published in 1949 with the title *Giant Brains or Machines That Think*.[13] Exactly thirty years later another book popular among scientists appeared with the title *Machines Who Think*.[14] One of the key questions debated at the 1984 conference mentioned earlier concerned the change in metaphor from "that" to "who."

Hopes and fears

A case history from the field of physics furnishes a good example of the ambivalence surrounding the future of thinking machines in science. The author of the history is Lew Kowarski, a physicist at Western Europe's largest nuclear research facility, Conseil Européen pour Recherches Nucléaires, in Geneva. Kowarski presented that history at an international conference on information processing held in Vienna in 1974.

Kowarski related how the use of computers in physics research facilities was at first regarded purely in terms of assisting human decision making, how the computer soon became the spearhead of a drive toward full automation, why plans to make the computer self-sufficient failed, and how this subsequently resulted in an accommodation between man and machine that he called "symbiotic." He thought the relationship had real possibilities but concluded with the uneasy feeling that it could be one of temporary convenience on the computer's part and that renewed attempts to achieve a complete automation, "to compress the human sector," as he put it, were, back in 1974, already in sight.[15]

The physicist's tale. In very abbreviated and simplified fashion, the tale goes like this. Research in high-energy physics involves observation of how the paths of elementary particles form tracks of drops, bubbles, or sparks when they pass through a specially prepared medium. Photographs of the tracks provide data crucial to physical theory; however, the work entailed in scanning thousands upon thousands of visual records, measuring each detail, and converting these measurements into meaningful physical information was, in the mid-1950s, fast

10. Herbert A. Simon, "The Steam Engine and the Computer: What Makes Technology Revolutionary," *EDUCOM Bulletin*, 22(1):2-5 (Spring 1987).

11. Pagels, ed., *Computer Culture*, p. 139.

12. Ibid., p. 151.

13. Edmund C. Berkeley, *Giant Brains or Machines That Think* (New York: John Wiley, 1949).

14. Pamela McCorduck, *Machines Who Think* (San Francisco: Freeman, 1979).

15. Lew Kowarski, "Man-Computer Symbiosis: Fears and Hopes," in *Human Choice and Computers*, ed. Enid Mumford and Harold Sackman (Amsterdam: North-Holland; New York: Elsevier, 1975), pp. 305-12.

making experimentation prohibitive.

The answer lay in computers—initially not for all the tasks involved, but only for what computers were then understood to do well, namely, to perform the number-crunching operations required for processing the measurements. Scanning continued to be entirely human, and a human operator controlled the robot optical sensor that took measurements of the tracks. As confidence in the computer's potential grew and as improved hardware and software came on the scene, however, plans were laid for total automation of the entire experimentation process. The human scientist would intervene only if the machine yelled for help.

The scheme did not work, and the fact that it did not may be reassuring of human superiority. But, Kowarski added, "the story is not ended: this kind of situation never stands still." He went on to say that improvements in computer-controlled image processing were already in view and could presage an era in which novel experimentation would be avoided because, being unmechanized, it would be regarded as too laborious. If this comes about,

> we will have adapted our demands to what can be supplied by a machine and not the other way around. A tool and a servant, even when blindly obeying its master's orders, has a subtle way of improving its own nature and putting limitations on the choice of orders to be issued.

Ambivalence revisited

Kowarski's uncertainties about the future persist with unabated intensity today. In a recent review of books currently on the market, Peter Denning, director of the Research Institute for Advanced Computer Science at the National Aeronautics and Space Administration's Ames Research Center, points up the continuing ambivalence surrounding computers as thinking machines—even in the minds of scientists who can lay valid claim to being authorities in the field.[16] According to Denning, of the three books he reviewed, one leaves the reader with the impression that the search for "machines that think" is futile; another is skeptical but not so sure that new computer architectures might not give the machine mindlike properties after all; and the third is confident that mindlike behavior can indeed be elicited from machines if their architecture can be made to resemble the human nervous system. To demonstrate confidence in this conception, the third book provides the details of a Cellular Model Arithmetic Computer, on which the author holds a patent.

Another pertinent review of some recency also reflects the continuing ambivalence surrounding the question of whether computers can actually think. This one was written by Raymond Kurzweil, who invented the world's first optical scanning system—the Kurzweil Reading Machine for the Blind—which converts printed text into synthetic speech. In his review, Kurzweil tends to debunk much of the enthusiasm propagated by AI researchers and asserts that the goal of the AI enterprise "should not be to copy human intelligence in the next generation of computers, but rather to concentrate on the unique strengths of machine intelligence, which for the foreseeable future will be quite different from the strengths of human intelligence." Nonetheless, Kurzweil's article concludes by noting that "some observers have actually suggested that artificial intelligence is inherently on the

16. Peter J. Denning, "The Science of Computing: Will Machines Ever Think?" *American Scientist,* 74:342-46 (July-Aug. 1986).

moving edge of technical feasibility, that it should be defined as those computer science problems we have not yet solved."[17]

LIVING-BRAIN MACHINES

A number of other recent writings suggest that Lew Kowarski's concept of a future characterized by man-machine symbiosis has evolved into the idea of a brave new world populated by immortal man-machine chimeras. One such work is a book composed largely of conversations with renowned computer scientists and their young protégés who go by the sobriquet "hackers."[18] The conversations took place in prominent AI laboratories in various parts of the Western world. Here we discover a sincere belief in, apparently accompanied by the beginnings of a serious attempt to produce, computer-controlled robots capable of performing brain surgery—not for the purpose of removing diseased brain tissue, but for the purpose of analyzing and simulating the neuronal and chemical makeup of that particular brain and downloading it—that is, transferring it, in information-age argot—to a man-made body.

The Tomorrow Makers was written by a journalist—albeit a science writer of respectability—and will no doubt be thought to suffer from its journalistic tenor. Not so Margulis and Sagan's *Microcosmos*, a work bearing all the tokens of scientific credibility.[19] Margulis and Sagan provide a bookload of scientifically based evidence to back up the proposition that machines undergo evolutionary adaptation no less than organic forms do and that the cell, the brain, and the steam engine are

> the technological innovations of chains of deoxyribonucleic acid (DNA), which have spent the past three billion years evolving vessels to carry themselves into new environments. Man and machine may both grow obsolete in the long run, as have 99.99 percent of all the species that have appeared. But it is equally possible that the two will merge, in some unimaginable form, to escape a dying planet.
>
> It may sound outrageous to suggest that life and nonlife could ever blend or breed, but viewed in the context of the history of life—from its prebiotic origins to its present day ability to manufacture itself in the laboratory—such a feat begins to look not only plausible but inevitable.[20]

A sober note

To return to the near-term future rather than end on a possibly depressing long-term note, the philosophy espoused by most of today's AI practitioners projects AI's value to science and humanity as coming from the growing collaboration between AI and research in cognitive science, wherein the goal is not the reification of human intelligence in computational forms but the use of computers as investigative devices for studying brain-behavior relationships. This use of computers has no substitute. Studying organs with billions of extremely small, sensitive, working parts packed together is totally new to science, and both ethical and technological limita-

17. Raymond Kurzweil, "What Is Artificial Intelligence Anyway?" *American Scientist*, 73:258-64 (May-June 1985).

18. Fjermedal, *Tomorrow Makers*.

19. Lynn Margulis and Dorion Sagan, "Strange Fruit on the Tree of Life," *Sciences*, 26:3, 38-45 (May-June 1986), adapted from Margulis and Sagan, *Microcosmos: Four Billion Years of Evolution from Our Microbial Ancestors* (New York: Summit Books, 1987).

20. Margulis and Sagan, "Strange Fruit," p. 40.

tions preclude investigations of the brain cells of higher animals while they are actually working and learning.

Cognitive science encompasses such areas as sensory information processing, visual perception, and speech recognition pursued by researchers identified with the neurosciences, cognitive psychology, computational linguistics, and the philosophy of science. These workers see AI as providing tools for research that can reduce ignorance about humankind and help pave the way for improvement in the human condition. Science itself may well pay the price of all this advancement, however, as these tools become increasingly autonomous and possibly mysterious elements in its processes.

Book Department

INTERNATIONAL RELATIONS AND POLITICS

BULKELEY, RIP and GRAHAM SPINARDI. *Space Weapons: Deterrence or Delusion?* Pp. xv, 378. Totowa, NJ: Barnes and Noble, 1986. $33.95. Paperbound, $11.95.

Bulkeley and Spinardi set out to offer a balanced survey of the possibilities and problems of any space-based defense of cities against nuclear attack, as proposed in President Reagan's Strategic Defense Initiative, and of all the related temptations of space warfare one might suspect may be connected with this.

They do largely succeed in producing a readable account, even though they pull together a very large amount of material and are meticulous about citing all their sources. At times one feels that they offer slightly too many quotations and citations as support for their conclusions. The prose is basically very readable, but the sheer volume of technical possibilities and alternatives that are discussed still occasionally overwhelm the reader.

The book's presentation is at times so fair-minded and matter-of-fact that the reader might fail to see an overarching viewpoint being presented, but Bulkeley and Spinardi indeed do have such a viewpoint. They doubt the technical possibility of a defense of civilians against nuclear attack; perhaps more important, they deduce from three decades of the history of such efforts that we should also distrust the commitment of either side, especially the United States, to any policy of mutual assured destruction.

The book is very valuable for all the detail it pulls together, dealing both with the overall bottom line of whether systems can work and with the narrow details of small parts of the systems. The account is also helpful on the gradations and kinds of antimissile defense that might be achieved, along with some of the military strategies to which these could be connected. Bulkeley and Spinardi may be slightly less at home in their conclusions about the complicated motives of the various American and other statesmen who have dealt with these issues, but the factual record put forward here is valuable to anyone trying to shape his or her own judgment.

GEORGE H. QUESTER

University of Maryland
College Park

CABLE, LARRY E. *Conflict of Myths: The Development of American Counterinsurgency Doctrine and the Vietnam War.* Pp. xiii, 307. New York: New York University Press, 1986. Distributed by Columbia University Press, New York. $30.00.

One of the latest books to examine American participation in Vietnam is Larry E. Cable's *Conflict of Myths.* This book has much to recommend it, but it also has some shortcomings.

The book can be divided into three sections. The first section discusses a number of wars in which guerrillas fought; there is a chapter on each war discussed. The second section has a chapter each on the doctrine concerning guerrilla wars of the United States Army, the Marines, and, in a combined chapter, the Navy and Air Force. The last section has chapters that deal with the communications between top levels of the armed forces, the State Department, and Washington, D.C., on what approach to take in Vietnam. This section deals only with the years between 1960 and the summer of 1965.

The basic argument of the book is that there are two kinds of guerrilla wars: partisan and insurgent. A partisan war is one where the "guerrillas operate as an auxiliary to the regular military forces of a nation. Partisans do not exist without external support, sponsorship, and control." On the other hand, "insurgents operate as armed political dissidents within a society seeking revolutionary social and political changes." The key difference between the types is that partisans are under external control with external supply lines while insurgences are indigenous operations—they do not depend on either external supply or control. Cable charges that the United States' Vietnam policy mistakenly viewed the Vietnam war as a partisan war, and that this mistake led to an erroneous strategy heavily dependent on closing the border and cutting the supply lines.

The wars covered in the first part of the book are "the Greek Civil War"; "South Korea, 1948-1954"; "the Philippines, 1946-1954"; "the Malayan emergency"; and "the banana wars, 1915-1934." It is difficult to understand why these wars were chosen in preference to others. For example, if Cable wanted to concentrate on American experiences, he could have added the Indian wars of the late nineteenth century or the post-Spanish American War experience in the Philippines. Or, if he wanted more up-to-date data, he could have looked at the French experience in either Vietnam or Algeria. All of these, and many others, have data that bear on his topic. Moreover, the chapter on the banana wars confined itself to Nicaragua and the Sandinistas. Why there instead of, say, Haiti?

It is easy to understand, however, why Cable chose Malaya. His description of the techniques employed by the British there seem to represent his ideal strategy. As painted by Cable, the following elements led to the success of the British: civilian control, with the armed forces subordinate to that control even on the armed forces' daily operations; elite units chasing guerrillas while less well trained units stood guard; good intelligence about the guerrillas; great knowledge of the local culture on the part of the authorities; and political steps to remove popular discontent. At numerous places in the subsequent discussion, the Malayan experience is used as a basis of comparison.

In the second section of the book where the discussion of the services doctrine occurs, the main emphasis is on the U.S. Army. Here he commits an error in charging that the army's doctrine is greatly influenced by Clausewitz and as a consequence overemphasizes the importance of a battle that destroys the enemy force. Clausewitz did emphasize winning battles, but his emphasis must be taken in the context of his complete argument. Clausewitz's main argument is that war is an extension of politics—wars are not fought to kill people but to achieve political objectives, and the conduct of a war must keep the political objectives in the forefront. Winning battles and killing enemy soldiers merely puts great pressure on the enemy to give in. The fact that the U.S. Army doctrine does emphasize the importance of battle is sometimes used to demonstrate its anti-Clausewitz bias.

The third section of the book shows the increasing involvement of the United States in Vietnam and ends with the actual employment of U.S. troops on a large scale. In it, he traces the position of a number of the major actors over time with an emphasis on their strategy recommendations.

Cable has done an impressive amount of research on the subject, consulting a number of archives as well as the usual secondary sources. To oversimplify, the first section is mostly based on secondary sources, while the second section stresses material from field manuals of the army. Cable has much more confidence than I do that field manuals represent the actual thinking of the army. The last section is heavily laden with archival material. The one type of source not used is interviews with the principles. That is too bad, for some of the actors in this drama might have been able to fill in lacunae in Cable's narrative.

This book has some interesting points for students of the Vietnam war, but it cannot be considered the definitive work on the subject.

O. ZELLER ROBERTSON, Jr.
Saginaw Valley State College
University Center
Michigan

DUNER, BERTIL. *Military Intervention in Civil Wars: The 1970s*. Pp. xiii, 197. New York: St. Martin's Press, 1985. $25.95.

This book, the produce of a project initiated at the Swedish Institute of International Affairs and completed at the Department of Peace and Conflict of the University of Uppsala in Sweden, studies the role of foreign military intervention in civil wars. It is divided into two distinct parts. Part 1 describes all the civil wars that ran their course in the decade of the 1970s. Twelve are recognized: Angola (1975-76), Burundi (1972), Cambodia (1970-75), Guatemala (1970-72), Iran (1978-79), Jordan (1970), Lebanon (1975-76), Nicaragua (1978-79), Pakistan (1971), the Philippines (1972-77), Rhodesia (1972-79), and Sri Lanka, or Ceylon (1971).

In a chapter called "The Polymorphy of Intervention," Duner examines the several facets of foreign intervention. One such facet concerns the propensity for intervention, measured by the number of acts of intervention, or by its intensity, that is, direct or indirect involvement in combat, in paracombat, or in support activities. A second facet highlights the instruments of intervention: invasion, arms supply, advisory service, armed blockade, monetary contribution, military training, and so forth. Still another presents a detailed correlation between the nature of intervention and the typology of interveners. Duner divides the latter into two main categories: interveners from industrialized or developed countries, and those from less developed or developing states. He then compares these two sets in terms of their levels of intervention, their frequency of involvement, and their geographical proximity to the countries affected by civil strife.

In part 2, entitled "Inter-Intervener Relations," Duner carefully probes the connection between states that intervene in the same civil war. He divides the interveners into three broad sets: the interveners by proxy—"the question is whether or not a state has been pressured into intervening by another state"; those who engage in it via military cooperation—"where intervention occurs with at least two agents assisting the same party, there is a very strong tendency to military cooperation between the two interveners"; and, third, those who jointly intervene because of shared economic and political ties in the international arena.

After finishing this book, I was left with the impression that I now knew everything there is to know about foreign intervention in civil wars, so thorough is its author and so meticulous and complete his analysis. Although his samples encompass only the civil wars of the 1970s, I suppose Duner intends his findings to be generalized to a broader range of similar phenomena. In that regard, I consider that the study's most interesting conclusion was the one that centrally challenged the established belief that industrialized countries, primarily, act to influence the course of civil strife. The statistics presented in this book argue otherwise, that the bulk of external participation originates with the less developed countries. These, it seems, dominate both in terms of acts of involvement and other criteria. Of the 50 interveners Duner enumerates, 34 are developing countries; and of 91 acts of involvement, 58 are conducted by members of the same group.

Two explanations are offered for this finding. First of all, as regards number, Duner commonsensically states that "the less developed countries are so numerous, they are bound to predominate" in the statistics. Second, because developing countries tend to cluster in discrete regions, when political matters in one developing country deteriorate into civil conflict, immediate neighbors are more ready to move in. Indeed, "about half to one third of all interventions refer to the actions of neighboring states."

A second interesting finding shatters an equally deeply held assumption of commentators on intervention, expressed in the usual dismissal of the proposition that Cuba, for example, or South Africa, could have independent motives for getting involved in Angola, or Vietnam in Cambodia, they being, of course, nothing more than proxies of the Soviet Union or the United States! Taking many long detours and offering some rather elaborate reasoning, Duner concludes, emphatically, that "it is impossible to demonstrate a single example of a state acting as a proxy for some other state." While granting the problem universal to all power research—that of trying to determine aims and intentions—he suggests that it makes more sense, in the area of intervention, simply to talk about "partner, allies, similarity of objectives, compatibility of interests or collaboration or coordination of activities."

In sum, it is quite impressive to see how, from a limited array of samples, Duner manages to draw a great number of plausible conclusions. I hasten to add that the monograph manifests a solid base of careful research, extensive documentation, and sound speculation. The bibliography, for example, lists no fewer than 150 books and 120 articles that Duner actually consulted, in addition to approximately 50 newspapers and magazines, including the *New York Times, Washington Post, Far Eastern Economic Review, Times* and *Time* magazine, all of which are, of course, quite indispensable to this type of research on contemporary topics.

TRUONG BUU LAM
University of Hawaii at Manoa
Honolulu

HARBUTT, FRASER J. *The Iron Curtain: Churchill, America, and the Origins of the Cold War.* Pp. xiv, 370. New York: Oxford University Press, 1986. $24.95.

In this well-crafted study of Winston Churchill and the roots of the cold war, Fraser Harbutt convincingly challenges the Americocentrism implicit in orthodox and revisionist interpretations of cold war origins. Rather than seeing the cold war emerge out of Truman's atomic diplomacy, or out of America's logical response to Soviet aggression in Eastern Europe, Harbutt traces its origins to Churchill's strategy for restructuring the postwar geopolitical arena and to a corresponding transformation of postwar public opinion.

Based on exhaustive research in American and European archives and in private manuscript collections, Harbutt's study examines the history of Churchill's thought and action regarding Soviet and American roles in world politics, from his first encounter with American culture in the late nineteenth century, to his campaign for Anglo-American intervention against the Bolsheviks in 1919, and finally to his eventful speech of 5 March 1946 at Westminster College in Fulton, Missouri. Harbutt argues that, after 1919, Churchill had a primary objective to create an Anglo-American alliance against Marxism and the Soviet Union. Most of Harbutt's book focuses on the political maneuverings of Churchill, Roosevelt, and Stalin between

1936 and 1946. World War II forced Churchill to repress his obsessive anti-Sovietism. Indeed, as a firm believer in traditional power politics, during the war Churchill engaged in bilateral talks with Stalin to refashion European spheres of influence. Meanwhile, Wilsonian moralist Franklin D. Roosevelt espoused an internationalist approach to assure postwar peace, and toward that end he favored accommodating Russia's postwar objective to secure friendly governments on its eastern and southern borders. Harbutt, however, also sees FDR as a shrewd geopolitician whose Declaration of a Liberated Europe at Yalta represented a cunning effort to check Stalin's ultimate ambition to subjugate Eastern Europe politically.

Harbutt is somewhat vague in pinpointing precisely the moment and circumstances surrounding the collapse of Big Three cooperation. As Stalin's plan to dominate Eastern Europe and Iran on the Northern Tier became clear, and as Stalin engaged Churchill in a war of nerves, Churchill resumed his effort to shore up England's weakened world position and shake America out of its accommodationist mode and into an Anglo-American alliance against communism. According to Harbutt, Truman and Secretary of State James F. Byrnes's late February 1946 shift in policy toward the Soviet Union from "accommodation" to "firmness" must be seen against the backdrop of Churchill's determined postwar effort to mold American public opinion and foreign policy. Churchill's speech in Fulton on 5 March 1946 signaled not only the dawn of the cold war, but the triumph of the former prime minister's crusade for British and American unity against Soviet communism. He also views America's firm opposition in the United Nations Security Council to Russian aims for military domination of Iran as an initial test of the new cold war geopolitical alignment. Well written and researched, Harbutt's book represents a fine effort to place cold war origins into a more historical geopolitical framework.

JOHN F. BAUMAN

California University
of Pennsylvania
California
Pennsylvania

MAYERS, DAVID ALLEN. *Cracking the Monolith: U.S. Policy against the Sino-Soviet Alliance, 1945-1955* Pp. xiii, 176. Baton Rouge: Louisiana State University Press, 1986. $22.50.

I can eliminate some of the suspense. This is a well-written, amply documented, and timely book. Those who are scholars will profit from the skillful use of historical materials to make a point. The general-interest reader—especially, though not necessarily, those interested in the 1950s—will find themselves intrigued by the twists, turns, and personalities associated with U.S. foreign policy. We all will profit from having our imaginations refreshed and stimulated about the interactions between domestic and international policymaking and rhetoric.

Policymakers in open political systems have to make policies, explain them to other policy players, and rally public support for their decisions. This often creates a dilemma: the preferred or best policy is not always the one that is either easiest to explain or for which it is easiest to mobilize support. A difficult policymaking situation is generally made worse by having created previously a rhetorical environment that skews attention away from reasonable discussions and toward emotional appeals. I believe Mayers, a professor at the University of California at Santa Cruz, properly finds that U.S. policymakers during the early 1950s were trapped by their overheated rhetoric and were thereby prevented from responsibly developing effective policies toward the People's Republic of China, especially, and the communist countries generally.

In the 1950s the belief in monolithic communism and too much reliance on military responses to essentially nonmilitary pressures combined with all sorts of diplomacy via wisecracks and verbal jingoism to inhibit U.S. policy. Ironically, while one should not suddenly believe that history repeats itself, Reagan's operation of foreign policy seems very like its 1950s predecessor. This déjà vu dimension makes Mayers's book important reading.

According to Mayers, with respect to the People's Republic of China during the 1950s,

U.S. policymakers simply ended up getting caught in the sticky web of their own rhetorical making. As he puts it,

> Although it is impossible to deny American policy makers of the 1949-1955 period some perspicuity and cleverness with regard to the Sino-Soviet alliance, despite formidable domestic constraints, it is essential to recognize the role of these leaders in creating the bugaboo of aggressive international communism that led to McCarthyism and greatly handicapped adroit diplomacy (p. 157).

Does this fit 1987 as well as 1954? Remember the "evil empire," "Libyan madman," the contras being referred to as "the moral equivalent" of our revolutionary forebears?

CARL F. PINKELE

Ohio Wesleyan University
Delaware
Ohio

MENON, RAJAN. *Soviet Power and the Third World.* Pp. ix, 261. New Haven, CT: Yale University Press, 1986. $20.00.

It may be that his editors have done Rajan Menon an unintentional disservice by selecting the title of this book, for his study is basically an interpretative essay on Soviet arms transfers to the Third World. At times one is made aware of fairly strong disagreement between Menon and other scholars regarding Soviet intents and capabilities. Menon writes from an ostensibly independent standpoint. I have found some of his remarks in passing to be often more provocative than his major theses. This is because in his longest chapters he introduces a great deal of very compressed history in support of analytical conclusions that are unveiled at intervals rather than presented in one or two meaty paragraphs.

Menon asserts that in 1917 Russia was not a mature capitalist society—pace Lenin; cites Soviet scholars counseling Third World countries against hasty measures in nationalizing their economies, thereby implicitly criticizing the nationalization measures of War Communism (1918-21); argues that "revolutionary democracies" in the Third World regard the state as the primary agent of modernization, summoning to my mind Stalin's immediate successor, G. Malenkov; suggests that caution in nationalization might provide a route toward realizing other social goals through the taxation of multinational corporations; points to the Soviets' new stress on profitability; and concludes that a striking resemblance exists between contemporary Soviet advice and the New Economic Policy of the 1920s. He calls this economic outlook "Bukharinist."

Menon underscores his assertion that Soviet cultural diplomacy in the Third World has been a gigantic failure. Nevertheless, the Third World provides the only theater in which the Soviet Union can act out its self-image—fantasy?—as a revolutionary agent for change.

Faute de mieux arms transfers constitute the key instrument for Soviet policy in these areas. According to the Stockholm International Peace Research Institute, the Soviet Union is the leading supplier—though generally not of the most advanced models—having overtaken and surpassed the United States as arms merchant. The two states provide 37 percent and 32 percent, respectively. Expatiating on the qualities of "frugality and pragmatism" that have characterized Soviet arms transfers, Menon indicates that the Soviets provide less than 3 percent of the total aid received by Third World countries from all sources—and that "revolutionary democracies" received only 35 percent of that sum. Since the mid-1970s the need to earn hard currency has accentuated Soviet motivation.

Menon devotes several worthwhile pages to reasoned disagreement with other scholars concerning the new emphasis in Soviet military publications on "power projection," which he views with greater caution than some other writers. He observes that the Soviet officer corps does not have a unified view on the relationship between local wars and escalation. Even if it is conceded that Soviet military power exceeds that of the United States, Soviet strategic analysis embodies the view that military power is only one component of the correlation of forces among states. While the Soviet Union might apply military force to regions close to the motherland, it is unlikely to do so in regions far removed in order to avoid the precedent of international humiliation on the scale of the Cuban missile crisis. Menon argues further that the lethargic Soviet economy militates against frequent attempts to "project power." The pages on Afghanistan, perhaps naturally, already seem outpaced by events.

In conclusion Menon touches on the idea that the Soviet leadership might resort to expansionism in order to divert its populations from the system's endemic problems. Under Gorbachev's leadership, however, the leadership has apparently resolved to travel another route. Menon's writing is clear, judicious, and sober. His argument might be better served if its architectonics had not been obscured by selectively draped evidence.

DALE LaBELLE

Somerville
Massachusetts

MULLINS, A. F. *Born Arming: Development and Military Power in New States.* Pp. viii, 147. Stanford, CA: Stanford University Press, 1987. $28.50.

DAVID, STEVEN R. *Third World Coups d'Etat and International Security.* Pp. xi, 91. Baltimore, MD: Johns Hopkins University Press, 1986. $22.50.

Why and how nations arm themselves are questions of great importance not only because of the obvious impact on interstate conflict, but also because of the role arms play in the internal process of development. One of the more controversial empirical findings in recent years is the discovery by the late Emile Benoit of Columbia University that military expenditure may not be detrimental to economic growth in developing countries. In his words:

> We have been unable to establish whether the net growth effects of defense expenditures have been positive or not. On the basis of all the evidence we suspect that they have been positive. . . . but we have not been able to prove this. Heavy defense expenditure does not, however, appear to have been associated with lower growth rates, even after adjusting for differences in foreign aid receipts and investment rates and this in itself is surprising (Emile Benoit, *Defense and Economic Growth in Developing Countries* [Lexington, MA: Lexington Books, 1973], p. 3).

The positive relationship found by Benoit between the defense burden and economic growth could, however, be spurious because economic growth could be caused by the inflow of other types of foreign resources, not just aid. There has also been considerable skepticism regarding Benoit's explanation that rising military expenditures stimulate private demand and encourage fuller utilization of production facilities. Several critics have argued that the main problem facing developing countries is not inadequate demand and underutilized capacity, but severe production bottlenecks in precisely those industrial sectors that are likely to be further strained to cope with additional defense demands.

It is in this context that A. F. Mullins's work makes a major contribution to our understanding of the defense-development debate. Concentrating on the 46 countries that have achieved independence in the last quarter-century, Mullins examines the relationship between military capability and political and economic development. In particular, Mullins examines a series of questions suggested by this relationship: what conditions promote or inhibit the acquisition of military capability? Do those conditions change with time or geographical circumstances? And most important, how does growth in military capability itself affect the rate of development?

Mullins finds that despite great variation in individual state performance in economic growth and the acquisition of military, it is possible to discern some general patterns. Military power tends to increase in the wake of economic progress rather than before it, unless the military power comes as a result of a patron's support. And of all the patrons of new states, the Soviet Union is the most generous when it comes to military assistance. Patron support can easily provide capability scores equal to the highest found in the data set, and so regimes interested in military power have an alternative to domestic resources in acquiring it. In general, those states that have done the best in gross national product growth appear to have paid less attention to military capability than others. This relationship is not perfect, but it holds right across the range from poor states to rich and from weak states to powerful. Those that did most poorly in gross national product growth not only paid more attention to military capability, but usually chose a patron's aid as the route to acquiring it. Whether the regime was military or civil, or came to power peacefully or via a revolution, had no obvious impact on the relationship between gross national product and military capability.

Most of the findings of empirical studies of this type are critically dependent on the definition of a few key variables and/or model specification. Benoit has been criticized at length on these grounds. Mullins's study centers on the development of a military capability index. This index in turn is used to show the adverse effects of militarization on development. Is this the appropriate measure of militarization, or would the more conventional budgetary categories of military expenditure be more appropriate for this task? In cases such as these, where the main finding is dependent on a somewhat arbitrary definition, Mullins should be encouraged to test the robustness of his important results through developing alternative yet theoretically sound measures of the military capability and/or defense burden.

Despite some doubts as to the generality of his results, Mullins's work presents a refreshing contrast to the usual knee-jerk subjective writings on the subject; his study represents perhaps the most sophisticated and objective condemnation of Third World militarization.

Whereas Mullins's study can be used to chide the superpowers on their irresponsible contribution to Third World militarization, David's excellent study does not necessarily condemn or condone superpower involvement in Third World coups. Instead, David argues that a significant number of Third World coups since 1945 have been deterred, suppressed, or supported by states seeking to protect friendly regimes or undermine hostile ones. For Third World leaders, manipulating coups in other countries can be a means of protecting their own country from outside aggression without overextending their military capabilities. For the superpowers, coups offer relatively risk-free opportunities for expanding influence without overt use of force.

David examines more than twenty coup attempts in detail. Among the cases receiving attention are those of Afghanistan, Angola, Chile, Ethiopia, Guatemala, Honduras, South Vietnam, South Yemen, Sudan, and Syria. David explains the nature and frequency of coups in the Third World and compares the experiences of the United States and the Soviet Union in attempting to exert control. These activities, he finds, reveal a critical area of superpower competition in the Third World that is often overlooked by scholars and policymakers.

Both David's and Mullins's books represent the culmination of serious objective research on one of the more vital areas of Third World studies—the impact of militarization on economic development and the projection of power. Together they suggest a K-Mart solution to third World geopolitics—do not waste money on arms; the true discount method of doing in one's enemies is to have them overthrown through low-cost coups.

ROBERT E. LOONEY

Naval Postgraduate School
Monterey
California

RASKIN, MARCUS. *The Common Good: Its Politics, Policies and Philosophies.* Pp. xii, 369. New York: Routledge & Kegan Paul, 1986. No price.

The ethics of social responsibility form the basis of Marcus Raskin's *Common Good: Its Politics, Policies and Philosophies.* Raskin contrasts the common good as a system of governance with the domination drives of the modern nation-state and the power of large-scale capital ownership. His vision presupposes a critique of militarism, cultural one-dimensionalism, and power brokering cloaked as marketplace economics, while his solutions emanate from the more benevolent side of human emotions, as "caring is essential to social reconstruction." His ideal is the parliamentary system without evil. Hence, the need for ethics, but also the still-present need for leaders and leadership.

Although the heart of Raskin's book lies not in its specifics, he spends considerable time on public policy, particularly the areas of the economy, military relations between nations, and social services. For Raskin, a reconstructed economy consists of three types of economic units: state-run enterprises, small businesses—namely, those with fewer than 500 employees—and worker cooperatives, with the state responsible for overall economic planning and the use of investment capital. The corporate world would disappear, but the market economy—along with its monetary and fiscal institutions—would continue to serve the purposes of social allocation. This would result in "greater public control by worker community associations, public boards and duly constituted governments in the federal system."

Behind these proposals lies the idea of greater community and individual involvement in social affairs, a theme that Raskin returns to repeatedly. Individuals, for instance, would reserve partial control to withhold tax funds from programs they do not favor. Similarly, local communities would nominate candidates for regional medical schools, and neighborhood courts would arbitrate local crime.

The common good is essentially an ethical program and philosophy for the policymakers. Raskin is less concerned with strategy and offers little to bridge the gap between the two strongest features of his book: program and future world. In theoretical terms, this world would no doubt appeal to the general population whose interests he so passionately advocates, but given Raskin's opposition to the present state of government, it is surprising that he should allow the shibboleth of participatory democracy to go untouched.

GARY ROTH

Jefferson Business School
Philadelphia
Pennsylvania

RUBENSTEIN, RICHARD E. *Alchemists of Revolution: Terrorism in the Modern World.* Pp. xxi, 266. New York: Basic Books, 1987. $17.95.

What this book does, better than any other extant, is relate terrorism to guerrilla warfare, nationalist, religious, or class ideology, politics, and revolution. Such linkages are established primarily through scholarly historical analysis, with synergistic effect flowing from the application of sociological psychological, and political science perspectives.

It is at once refreshing and productive of insight to find a work encompassing not only those groups most frequently labeled—from the official American perspective—terrorist but also such disparate and close-to-home groups as abolitionist John Brown's, the Ku Klux Klan, the Nicaraguan contras, the American/South Vietnam Phoenix teams, and, arguably, the Boston Tea Party participants. Within Rubenstein's analytical framework—more Leninist, perhaps, than the Marxist it is acknowledged by Rubenstein to be—these seem to evince surprising commonalities. Closer at hand are reasonable explanation and application of the dictum that one person's terrorist is another's freedom fighter.

Eschewing the media-inspired supposed wisdom that terrorists are either hysterical fanatics or mindless automatons manipulated by states sponsoring terrorism for their own purposes, Rubenstein focuses upon the disaffection of the intelligentsia as a prime cause of terrorism; it is these who are the would-be alchemists of revolution. Such disaffection tends to reach terroristic fruition, with the essentially conservative logic of terrorism reaching conclusion, when four conditions are present: postwar disorder, nationalist solidarity, modesty of aims, and opposition weakness.

Critically juxtaposing the successes of Communist revolutions in Russia, China, and Cuba with each other and with struggles for national liberation, Rubenstein offers advice to the West on dealing with contemporary terrorism. While there are those, including myself, who would take issue with his characterization of American interests as imperial, he is correct that terrorism will not disappear simply because the West ignores it or seeks to destroy it militarily. Even those who do not consider the United States an oppressor nation, then, may find provocative Rubenstein's call for the young intelligentsia to be allowed to reunite through collective action with the people of whom they are the best and the brightest. Provocative, indeed.

DENNIS D. MURPHY

Armstrong State College
Savannah
Georgia

VINING, JOSEPH. *The Authoritative and the Authoritarian.* Pp. xix, 261. Chicago: University of Chicago Press, 1986. $25.00.

Joseph Vining's eclectic work attempts to delve into the sources of legal authority. It is a humanistic view of the relationship of law to theology and the dilemma of the lawyer. Quoting from Frazier's *Golden Bough*, his introduction views human thought as having progressed from magic to religion to science in its attempt to give order and meaning to our world. The law, however, never became scientific. The law is a hierarchical way of ordering our thoughts that is nonmathematical. Interestingly, the law that has not become scientific may nevertheless teach the sciences valuable insights because unlike science it "leaves nothing out." As such, it is closer to philosophy than the narrow concerns of some of the empirically guided fields of thought.

The central problem of Vining's work is the difference between the authoritative and the authoritarian. To one, obedience and respect are willingly granted because of the authority's superior knowledge or some other attribute that is highly regarded. The authoritarian is obeyed not willingly but perhaps out of fear or simple power relationships. In order to distinguish between the two, Vining examines the legal method, institutional arrangements, and the lawyer's professional life. Theology, "the sister discipline of law," is viewed as an underused source of assistance in grappling with the problem of authoritative and authoritarian rule making. The framework for his analysis revolves around the bureaucratization of the court system in the United States. A good deal of attention is spent analyzing the implications of eliminating the Supreme Court as a source of law. This is meant to be an exercise aimed at learning more about the legal method, the relationship between the legal method and institutional structure, and the place of hierarchy in legal thought. The volume also examines the other two "grand institutions of law," the legislature and the electorate. This involves inquiries into such questions as, Why do some of us feel more comfortable with the authority of a legislature rather than the commands of angels as revealed in holy books? Or does majority rule necessarily supply legitimacy to rule making? Are decisions in nondemocratic systems necessarily illegitimate? After a brief discussion of what assistance the legal profession may hope to obtain from other disciplines in attempting to answer some of these questions, Vining closes with the idea that the Supreme Court may well be replaced by its surrogate, an impersonal system and process.

The volume, while extremely well written and thought out, at times dwindles into an examination of esoteric assertions. At times it is difficult to follow. Nevertheless, it is highly recommended reading to a wide audience from philosophers and theologians to attorneys and humanists.

JOHN ROBEY

University of Texas
Austin

AFRICA, ASIA, AND LATIN AMERICA

BIX, HERBERT P. *Peasant Protest in Japan, 1590-1884.* Pp. xxxix, 296. New Haven, CT: Yale University Press, 1986. $30.00.

WALTHALL, ANNE. *Social Protest and Popular Culture in Eighteenth-Century Japan.* Pp. xviii, 268. Tucson: University of Arizona Press, 1986. $19.50.

Most English-language studies of premodern Japanese society that were produced during the past number of decades by and large disregard the vast majority of the Japanese, namely, the common people who constituted upward of 80 percent of the population over the centuries. Curiously, within just the past year or so, five books that redress that imbalance have been published. Two of those books, which are under review here, deal with the peasants and townspeople of the Tokugawa period, 1600-1868, in their struggles with the ruling (samurai) class, which constituted 8 to 10 percent of the population.

Both Bix and Walthall portray in moving detail the plight of the peasant classes in Tokugawa feudal society, and they examine the reasons for peasant protests, the various types of protests that were mounted, the beliefs and ideals that inspired the common people as they strove to better their lot, and the results of those efforts. Readers will be struck by the wretchedness of the conditions under which most people lived under the Tokugawa shogunate: tied to the land, the peasants were demilitarized, depoliticized, and oppressed economically, socially, and ideologically. Subject to shogunal oppression in the forms of corvée service and the expropriation of their economic surpluses by way of assorted taxes that commonly siphoned off 50 to 60 percent of the yield of the land, the masses endured additional suffering at the hands of local village and town officials, rich landlords, merchants, and others.

Under those intolerable conditions unrest was inevitable. In order to remedy their situation, peasants engaged in various forms of protest ranging from polite petitions for the reduction of taxes to violent uprisings involving anywhere from a few dozen to as many as 200,000 people. The Tokugawa regime was shaken by some 3000 such uprisings, and by the end of that period peasant unrest had escalated to such a level that it was more responsible than any other factor for the collapse of the shogunate.

In their analyses of the causes, characteristics, and effects of peasant protests, Bix and Walthall bring to light the complexity and sophistication of peasant opposition to their oppressors. We see how peasants manipulated the language of Tokugawa officialdom to have it serve their ends; how they developed their own views on life and society in distinction from the ideologies imposed from above—mainly elitist, conservative neo-Confucianism; how peasant beliefs and values called into question the legitimacy of the political order and manifested the peasants' recognition of the finally arbitrary, not *de nature,* character of shogunal authority; and how the peasants used certain religious symbols, especially those of the always somewhat renegade mountain shamans, to give a sacred character to their protests.

Although these two books are similar in their subject matter, Bix and Walthall adopt different approaches in their studies of Tokugawa society. Bix is determinedly Marxist in his approach, and although some readers might be a bit put off by his sometimes strident declarations, his highly erudite book is representative of the best modern Japanese scholarship on the subject. Walthall's book, which is less clear and precise in its approach than Bix's, is so strongly colored by the author's overwhelming sympathy for her peasant subjects that her analysis of peasant protests tends to degenerate at times to a good-guys-versus-bad-guys level. Also, Walthall's work is annoyingly repetitious in parts as she makes the same points—for example, that peasants drew on their everyday experiences in formulating petitions—again and again. On the other hand, Walthall gives the reader a better feel for the details of peasant life in Tokugawa than does Bix, and her lengthy material on the documentary sources for the study of peasant protests is especially valuable. Whereas Bix traces chronologically the development of peasant protest movements from the late sixteenth century through Japan's transition to capitalism in the Meiji period, which spanned the years 1868-1912, Walthall concentrates mainly on the latter part of the eighteenth century and presents a more thematic treatment of peasant protests. Both Bix and Walthall offer comparisons between peasant struggles in Japan and those in other, European, countries. Thus the books complement and supplement each other, and both serve as antidotes to much modern Western scholarship on the Tokugawa period.

Now we need similar studies of early and medieval Japanese society.

NEIL McMULLIN

University of Toronto
Mississanga
Ontario
Canada

CHANDA, NAYAN. *Brother Enemy: The War after the War*. Pp. xiv, 479. New York: Harcourt Brace Jovanovich, 1986. $24.95.

SHAPLEN, ROBERT. *Bitter Victory*. Pp. vii, 309. New York: Harper & Row, 1986. $16.95.

It has been over a decade since the end of American involvement in Indochina and the reunification of Vietnam, with the result that most Americans have turned their attention elsewhere. There has not been an end to warfare in Indochina, however, and this tragic aftermath has become grist for the mills of two of the most knowledgeable journalists to have covered events in Southeast Asia. Their efforts invite comparison because they both purport to cover essentially the same ground, that is, the ten years following the reunification of Vietnam in 1975.

Nayan Chanda, chief of the Washington Bureau of the *Far Eastern Economic Review*, has produced in his *Brother Enemy* a solid and detailed political history of the diplomatic and military entanglements that enveloped Southeast Asia in the late 1970s. He argues that Vietnam attempted to concentrate on reconstruction at home and independence in its foreign policy within a framework of regional security. But Vietnam's idea of regional security was challenged by Cambodia, which pressed irridentist claims, and by China, which traditionally preferred a multistate system on its southern border. These security interests drew the radical and genocidal Pol Pot regime into an uneasy alliance with the increasingly moderate Chinese regime. On the other side, Vietnam allied with the Soviet Union, discriminated against ethnic Chinese, and finally, in 1979, invaded Cambodia.

During these diplomatic maneuvers, the United States played a role that manifested itself as the outcome of a policy dispute within the foreign policy sector of the government. Early in the Carter presidency, Secretary of State Vance had pushed for normalization of relations with Vietnam, but this effort foundered on Vietnamese insistence on reconstruction aid, the growing exodus of boat people, and the recalcitrance of Congress. Carter was more persuaded by National Security Adviser Brzezinski, who sought China's partnership in an anti-Soviet alliance. The way to obtain this was to normalize relations with China rather than Vietnam. As a result, the United States supported Cambodia against Vietnam and, when China invaded Vietnam in 1979, the United States gave its silent approval.

While written with the flair of a journalist, *Brother Enemy* is also solidly grounded in political history and a passing acquaintance with scholarly secondary sources. Despite its subtitle, *A History of Indochina since the Fall of Saigon,* the coverage effectively stops in 1980.

For readers of the *New Yorker* who are accustomed to the contributions by Robert Shaplen, his *Bitter Victory* will cover familiar territory. Indeed, much of the book has appeared in serial form in the magazine. A veteran reporter for over four decades in Indochina, Shaplen returned in 1984 to record his impressions, resulting in a book that is less a detailed political history than the Chanda book and more of an impressionistic contribution to our knowledge of postwar Indochina.

Because Shaplen's book is based upon a recent trip, it is somewhat more focused upon the events of the early and mid 1980s. His focus is also more on domestic political and economic trends and less on foreign affairs. In a range of interviews from those with the leadership on down to the man in the street, he chronicles the failure of Vietnam's economic initiatives of the late 1970s and the more flexible economic experimentation being done as a result. He also notes a reformist trend in the political realm, where older and more doctrinaire leaders have given way to younger pragmatists who have much in common with their Chinese counterparts.

To conclude, both books are similar in that they were both written by journalists and concentrate on the same area. Both authors also employ an oxymoron in their titles to highlight the almost schizophrenic twists that this history has taken. Both agree that the key to national reconciliation in Cambodia lies with Prince Norodom Sihanouk. They differ in that the Chanda volume emphasizes the period from 1975 to 1980 while the Shaplen volume emphasizes the first half of the 1980s. They differ also in that Chanda's volume is a much more tightly written political history that focuses on the diplomatic actors in leadership positions, while the Shaplen volume, by contrast, is somewhat more impressionistic with a greater emphasis on domestic issues in Vietnam and how they affect Vietnamese society. Both can be read with profit. Together they tell us much about the turbulent decade since 1975 in Indochina.

WAYNE PATTERSON

Vanderbilt University
Nashville
Tennessee

DAY, ARTHUR R. *East Bank/West Bank: Jordan and the Prospects for Peace.* Pp. ix, 165. New York: Council on Foreign Relations, 1986. $17.50. Paperbound, $8.95.

GHARAIBEH, FAWZI A. *The Economies of the West Bank and Gaza Strip.* Pp. xvi, 182. Boulder, CO: Westview Press, 1985. Paperbound, $16.50.

In a concise and well-written survey, Arthur Day provides a thorough introduction to contemporary Jordan. *East Bank/West Bank* opens with the basic political and diplomatic history of the country. In subsequent chapters he reviews the condition of the monarchy, the threat that various social conflicts represent to the cohesion of Jordanian society, the role of the army, and the status of the economy and of Jordanian-Palestinian relations. In a concluding chapter, Day turns to U.S.-Jordanian relations and suggests that a more consistent and sympathetic U.S. policy toward Jordan would advance the Arab-Israeli peace process, help secure the position of the monarchy, and generally advance U.S. interests in the region.

Though Day takes quite a sympathetic view of his subject, he makes very clear the impressive array of problems that Jordan faces. The monarchy appears as an institution whose legitimacy is fragile, held together by the force of King Hussein's personality though constantly buffeted by pressure from both a middle class seeking democratization and Islamic critics of the regime. In these circumstances the future of the throne is seen as tenuous, even though the succession procedure is relatively clear.

The rift between East Bankers and West Bankers, who have gradually emerged as a majority, emphasizes the problems surrounding efforts to build a cohesive national identity, especially when the successful integration of former Palestinians into Jordanian society apparently requires the abdication of Palestinian national aspirations that remain very much alive. Beyond this, Jordan's status as a small military power bordering the most powerful states in the region, and the dependence of Jordan's economy on regional and international trends are two additional reasons to be concerned about Jordan's future. Overall, Day offers the image of a state that has done unusually well simply by muddling through. But Jordan's intractable problems—the fragility of its political and economic institutions, the delicate balance of social forces, and a precarious regional situation—suggest hard times ahead.

Fawzi Gharaibeh presents a straightforward, descriptive economic survey of the West Bank and Gaza Strip since 1967, when Israel occupied those territories. Recounting briefly the development of these economies, he then offers a detailed sector-by-sector survey, using available statistics to illustrate the status since 1967 of labor force composition, agriculture, industry, services, construction, and trade. Two concluding chapters summarize the effect of Israeli policies on the overall economy of the territories and offer Gharaibeh's views on the economic future of the territories.

Gharaibeh is not polemical, though the status of the work as a critique of the occupation and its economic consequences is clear. Gharaibeh generally makes careful use of his material, and his analyses of the data are straightforward. In this sense, the book offers few surprises: the internal economies of the West Bank and Gaza Strip were generally underdeveloped prior to 1967, and the experience of Israeli occupation has further eroded the potential for local economic growth. Industry remains rudimentary, service sectors are relatively small, agricultural production is tenuous, tourism benefiting the Arab sector has decreased, investment is low, employment opportunities are limited, especially for skilled workers, and emigration is high. Gharaibeh shows how the economies of the territories have become linked to Israel, in terms of labor, trade, and industry, and how this, together with explicit Israeli policy, has further impaired the local economies. Whatever the future of the territories, Gharaibeh's measured and somber assessment seems justified in light of the tremendous obstacles to economic growth.

STEVEN HEYDEMANN

University of Chicago
Illinois

DORE, RONALD. *Flexible Rigidities: Industrial Policy and Structural Adjustment in the Japanese Economy 1970-1980.* Pp. vii, 278. Stanford, CA: Stanford University Press, 1986. $32.50.

SAXONHOUSE, GARY R., and KOZO YAMAMURA, eds. *Law and Trade Issues of the Japanese Economy: American and Japanese Perspectives.* Pp. xx, 290. Seattle: University of Washington Press, 1986. $40.00.

Both of these books deal directly with two aspects of the Japanese economic challenge. The Dore book describes how Japan not only survived the economic shocks of the 1970s, including the two oil shocks, but actually managed to adjust its economy in order to prosper in the 1980s. It recounts the measures taken by both business and government to protect and enhance the Japanese economy. The Saxonhouse and Yamamura book, on the other hand, focuses on the problems associated with these decisions and other long-established aspects of the Japanese economy for foreign economic interests. Thus these two books complement each other nicely. One describes what has happened in terms of a success story and the other describes some of the intended and unintended negative international consequences of the same decisions.

Dore's *Flexible Rigidities* has as its central thesis the argument that while the Japanese economic system may appear to many outsiders as excessively rigid, it is actually much more flexible than its Western competitors in adjusting to rapidly changing economic situations. Japanese labor practices are central to this understanding of Japanese flexibility, and Dore offers interesting and useful data to support his case. Even the casual reader will find interesting Dore's discussion of the artificially low unemployment rates and his attempt to determine who bore the brunt of the adjustments within Japan to the oil shocks. For those unfamiliar with recent Japanese economic history, Dore offers two chapters of background data on the economic challenges of the 1970s and how Japan's economy changed during that decade. He then proceeds to detailed analyses of how the combined responses of Japanese corporations, workers, and government proved to set the stage for Japan to become the world's most successful economy by the mid-1980s. Finally, the various themes are brought together and well illustrated in a three-chapter discussion on how the Japanese textile industry dealt with its economic problems. This is an excellent and well-documented book for any reader who is seeking to understand why the Japanese are doing so well in the 1980s while other major economies are having a more difficult time recovering from recent recession and foreign trade threats.

The Saxonhouse and Yamamura book is sponsored by the Japan-US Friendship Commission and the Japan Foundation, and its purpose is to promote understanding in both countries of the impact of laws and governmental practices on the conduct of international trade and commerce. Its style is much more detailed and legalistic—and less enjoyable to read—than the Dore volume. Much of the book focuses on such complicated issues as Japanese laws pertaining to consumer finance, antitrust policy, industrial policy, product liability laws, administrative guidance, foreign exchange and fair trade, patents, character merchandising, and franchising. In addition, useful legal case studies are presented dealing with the protection of foreign computer software, savings deposits in Japan, and U.S.-Japanese trade friction issues regarding automobiles, televisions, steel, and machine tools.

While the Dore book will be useful reading for a wide range of readers interested in the Japanese phenomenon, the Saxonhouse and Yamamura book will probably be of great interest to a much smaller group of specialists involved in comparative legal system studies, international trade conflict research, or businesspersons interested in doing business in Japan.

RONALD J. HREBENAR

University of Utah
Salt Lake City

FAIRBANK, JOHN KING. *The Great Chinese Revolution: 1800-1985*. Pp. xiii, 396. New York: Harper & Row, 1986. $20.95.

Someone should connect past and present, specifically the late imperial China of the nineteenth century with the Chinese Republic after 1911 and the People's Republic since 1949. Many books on these two centuries are now available, published mainly in the last forty years, but to be scholarly one has to specialize and leave the broad view to be cobbled together by textbook writers, popularizers, and similar types who are often least qualified to do it. What we need is an ex-professor who is not up for tenure, and who doesn't care about reputation. On that basis the integration of China's past and present can be a lot of fun (p. ix).

I would not say that John Fairbank's most recent book is exactly fun, but it is a real pleasure to read. The contents may be scholarly, but it makes no pretense of being a scholarly book. Except for quotations, it has no footnotes. Nor is it properly a textbook, because it has no lengthy bibliographies to add to its cost. Essentially, it is a book for reading, written by the undisputed dean of modern Chinese historians. It is also a book that clearly has something to say.

It is not easy being a dean. In the 1950s there were the right-wing McCarthyites who suspected anyone showing an interest in China of being a Communist agent. Then in the 1960s came the left-wing revisionists for whom anyone over thirty must be an agent of imperialism. Fortunately, these people soon became over thirty themselves, but they have been replaced by the new group of revisionists, who while less vociferous in their language have still directed their criticism primarily toward what they call the "China's response to the West" approach of John Fairbank and its implications of great Western chauvinism. It is no wonder, then, that the dean might like to set some of the record straight.

Thus, in chapter 1, "Understanding China's Revolution," and especially in chapter 3, "Some Theoretical Approaches," Fairbank discusses a number of approaches to modern Chinese history, criticizing both the comparative sociological and the response-to-the-West lines. He stresses, as he always has, that modern Chinese history has been shaped by an extremely complex mix of forces including tensions between new growth and old institutions, the impact of domestic disorders and foreign troubles, and the competition between modernization of material-intellectual life and the more slowly developing social change of values and institutions. At the same time, he seeks to "correct the gross imbalance by which the center of gravity of China's history has seemed to lie in the hands of foreigners outside the country," maintaining that disaster came to China in the nineteenth century because China, unlike Japan, was so unresponsive to the pull of Western gravity. According to Fairbank, China's great revolution of the twentieth century began with the disasters of the nineteenth, the roots of which lie in the eighteenth century or even much earlier ones. It is not so much, then, the introduction of the new but the struggle to transform the old that constitutes the major theme of this book.

It is in writing about the old that John Fairbank truly excels. His brief description of Ching political and social institutions in chapter 2 is a masterpiece of writing. If I were to make any criticism of this book, it is simply that I sometimes feel Fairbank attempts to see too much of present-day China in terms of its past. For example, it is easy to blame tradition for much of the bureaucratic inertia that plagues the People's Republic, but it should be kept in mind that bureaucracy also seems to be an inevitable by-product of modernization itself. One need only to look at what has happened in our own universities, not to mention our government in Washington, to see that one hardly needs a tradition of Confucian bureaucrats to suffer from this particular malady. How far one can push the influence of tradition is often highly debatable. This is a marvelous book, however, highly informative, sound in its judgments, wise in its predictions, and a joy to read.

W. ALLYN RICKETT

University of Pennsylvania
Philadelphia

GRUNDY, KENNETH W. *The Militarization of South African Politics*. Pp. ix, 133. Bloomington: Indiana University Press, 1986. $18.95.

Traditional democratic theory insists on a depoliticized military combined with civilian authority's clear control over strategic issues. How fully the barracks are under the command of an elected government and parliament is considered to determine the division between democratic and military rule. In white South Africa, this distinction does not apply. Increasingly, the entire society is becoming militarized. The boundaries between politicians and the military are fluid: personnel are interchangeable, and the two spheres merge. In Latin America the armed forces are usually politicized; in the beleaguered ethnic state, however, the polity becomes militarized. The political power of the State Security Council, in which military influence is strong, epitomizes this trend together with the spread of Joint Management Committees, a kind of parallel informal bureaucracy.

Kenneth Grundy documents these trends convincingly. Together with Philip Frankel's work on the South African military-civilian relationship, *Pretoria's Praetorians,* Grundy's earlier study on the black component of the South African security apparatus laid the groundwork for a better understanding of a neglected theme difficult to research. Grundy's main thesis that "the security establishment enjoys a greatly enhanced role in policymaking," however, is hardly original or surprising, given the development of the South African crisis. More than another catalog of acts of destabilization in neighboring states, one would like to know about the conflicts and competitions between Pretoria's militaristic powerholders. Above all, will the military eventually take over?

A suspension of white democracy and the establishment of a reformist military dictatorship are often mentioned as an alternative to black and white extremism. Great obstacles bar fulfillment of this technocratic vision, however. First, unlike the military in Latin America, the South African military does not constitute a separate caste; 90 percent are conscripts. These draftees reflect the divisions in white society. Their morale, already questioned by the army's use in the civil unrest, would be further undermined if they were called upon to serve a military dictatorship. The fewer than 18,000 professional soldiers could not run an advanced industrial society against widespread resistance. Second, the South African military is still, at least in theory, steeped in the British tradition of loyally serving the civilian government in power. That does not preclude that senior officers are politicized and take an active part in intergovernmental rivalries. It is well known that high-ranking members of the South African Defence Force frequently counteracted the Department of Foreign Affairs' policy of peaceful neocolonial relationships with neighboring states by actively working to destabilize "Marxist" governments. South Africa's political isolation has above all cut off its military personnel from foreign training and posting opportunities. That, in turn, has led to a parochial and regional outlook at the expense of realistic, global assessments by many senior personnel. Third, the close working relationship between the military and the ruling National Party that consolidated with the ascendancy of former Defence Minister P. W. Botha has made the military an integral part of government decision making, so that there is no need to take it over formally. Unlike the situation in several states that have experienced military takeovers, in South Africa the army leadership is on the same ideological wavelength as the government.

KOGILA MOODLEY

University of British Columbia
Vancouver
Canada

HOSTON, GERMAINE A. *Marxism and the Crisis of Development in Prewar Japan.* Pp. xviii, 401. Princeton, NJ: Princeton University Press, 1986. $40.00. Paperbound, $15.00.

In J. Grump's *Origins of Socialist Thought in Japan* (New York: St. Martin's Press, 1983), a Japanese syndicalist is quoted as saying, "The history of the world over the past few thousand years was precisely the history of the struggle for bread between the working class and the ruling class."

As this quotation would suggest, the formal terminology of Marxists, scholars, and practitioners can often be subject to imprecision and looseness of use. It is in this spirit that *Marxism and the Crisis of Development in Prewar Japan,* based on extensive use of Japanese sources, should be read. The work explores a variety of forms of an always-topical intellectual debate in Japan on the nature of capitalism. The debate covers the period of 1927 to the recent decade and focuses, in the main, on the controversies initiated by two broad schools of Marxist discourse. The Rōnō-ha, a labor-farmer group that split from the Japanese Communist Party, and the Kōza, or "feudal" core, so named because of its emphases on feudal remnants in Japanese economic enterprises, held specific positions about the quality, direction, and developmental trends in Japan's history. In essence, these topics evolve around forms of Japanese national values long since muted in public discussions, namely, the imperial—emperor—system and the significance to the Japanese people and state of the Kokutai theories of ideology.

All of the chapters in this text have a strong academic and even esoteric flavor because different parts stress the arcane Rōnō-ha versus Kōza arguments in evaluating the changing economic successes and failures of twentieth-century Japan. Because Hoston, a professor at the Johns Hopkins University, deals with a half-century of philosophical exchanges, the reader is required to remember considerable subtle information in moving from the general to the particular in dealing with Marxism in its differentiated forms. These forms have many expressions, ranging from the avowed purposes of some Marxist scholars to "promote a leftist revolution against what they perceived to be tremendous social and political acts resulting from Japan's rapid capitalist development" to intellectual dissection of the postwar economic regimes and the moral tenor of "Japan, Inc." In the earlier instance the traditional roles of the Emperor, the Privy Council, and the Kokutai are at issue. More contemporaneously, the "current situation" is commented on within a different and modified Marxist framework. Within the fifty-year span covered by this work, the reader will learn, not surprisingly, that careful Marxist theorists have demonstrated a patient willingness in adapting orthodox positions to conform to mass Japanese public values. Hoston weaves into her narrative the complex shifts of nuances and intellectual clashes of Communist Party policy, Communist International directives and theses, and the specific requirements of the Supreme Command Allied Powers' occupation.

There are no overriding conclusions in *Marxism,* a particular pity because much theoretical potential could have been drawn for the Marxist approaches in distinguishing between the "political" and "economic" determinants in present appreciations of Japanese capitalism. Neither Rōnō-ha nor Kōza-ha can currently offer a serviceable and sound explanation for Japan's perennial problem in maintaining access to industrial resources and materials, the international market, and in ameliorating Japan's tense relationships with its trading partners of the industrialized world. The Marxist scholars, mentioned throughout the piece, are on safe enough ground in documenting and stressing Japanese economic difficulties in historical context, particularly when the discussion elaborates on the significance of Japanese imperialism and the rise and partial demise of Japanese fascism. They are less able to apply their disciplined intellectual efforts and their philosophical assumptions to a Japan that has had land reforms and that is both modern and "democratic." Nevertheless, this book dissects much information pertaining to left-wing critiques designed to enhance Japanese capitalism in support of state power. But can these discussions and explanations be considered as "suggestive of analogous pattern[s] of development emerging in the Third World today?" Hoston poses the question early on. But this book draws on commentaries by Japanese Marxist theorists, intellectuals, and leaders. It is provocative, indeed, to assume that similar lessons and conclusions can be drawn from non-Japanese patterns of national development.

RENE PERITZ

Waseda University
Tokyo
Japan

KLEIN, HERBERT S. *African Slavery in Latin America and the Caribbean.* Pp.

311. New York: Oxford University Press, 1986. $22.95.

SLATTA, RICHARD W., ed. *Bandidos: The Varieties of Latin American Banditry.* Pp. viii, 218. Westport, CT: Greenwood Press, 1987. $37.50.

Fifteen years ago, Eric Hobsbawm urged fellow social historians "to watch what we are doing, to generalize it, and to correct it in the light of the problems arising out of further practice." Richard Slatta and Herbert Klein have both followed this injunction, each led by the distinct requirements of his particular area of social history: Slatta toward exploration, Klein toward synthesis.

Bandidos is concerned with correcting the theoretical foundations of a new field of social history by examining Hobsbawm's influential concept of "social banditry" in the Latin American context. Slatta and his collaborators study nineteenth-century and twentieth-century instances of banditry in Argentina, Bolivia, Brazil, Colombia, Cuba, Mexico, and Venezuela. Two short chapters look at Latin American banditry in Hollywood cinema and in criminological theory.

Through detailed use of judicial and police archives as well as sources from folklore and literature, the authors of *Bandidos* question Hobsbawm's view of banditry as an archaic form of social conflict in which the outlaw champions the interests of the downtrodden. Slatta notes that "the close ties of class and camaraderie that theoretically bind social bandits and peasants together do not surface in the Latin American context."

The chapters in *Bandidos* depict patterns of violence that developed in complex ways from an absence of economic opportunity. Banditry could become a way of life in remote frontier regions or in times of political instability. While some actions of social banditry did occur, bandits usually sought their own economic and political interests rather than those of the peasantry as a whole. They readily cooperated with powerful social elites and, at times, were even willing to switch sides to become law enforcement officials. *Bandidos* concludes that Latin American outlaws seldom deserved the nobly romantic images given them by folktales and literature. They were rebels driven by private questions of wealth and vengeance, not incipient social revolutionaries.

Klein's book reflects his command of the extensive and highly dynamic academic literature on slavery, offering a lucid comparative survey of African slavery in Portuguese, Spanish, and French-speaking America. Klein first examines the chronological development of slave labor systems from the sixteenth through the nineteenth centuries in parts of Latin America and the Caribbean where European conquest and capitalism created a demand for labor unsatisfied by European lower classes or Indian peasants.

He provides an able analysis of how the inhumane characteristics of slavery sustained the sugar plantation systems of Brazil and the Caribbean for over three centuries under shifting economic and political conditions. Klein notes the close association of slave labor systems with gold and diamond mining and with export cash crops like sugar, cacao, and coffee. At the same time, he depicts the lesser-known roles of slavery in urban craft production, frontier and coastal transport, truck farming, cattle raising, and fishing, stressing the importance of regional variations.

In the second half of the book, Klein abandons a chronological framework in order to discuss in a thematic fashion how slaves created their cultures and communities in Latin America and the Caribbean. He looks at the impact of the slave trade on demographic patterns; at the diverse forms of slave resistance and rebellion; at the social conditions of free blacks and mulattoes; and at the influence of different national patterns of abolition.

Klein's cogent synthesis and Slatta's exploratory questioning both support Hobsbawm's assertion that "it is a good moment to be a social historian."

ARTHUR SCHMIDT

Temple University
Philadelphia
Pennsylvania

PERETZ, DON. *The West Bank: History, Politics, Society, and Economy*. Pp. xi, 173. Boulder, CO: Westview Press, 1986. $22.00.

MISHAL, SHAUL. *The PLO under Arafat: Between Gun and Olive Branch*. Pp. xvi, 190. New Haven, CT: Yale University Press, 1986. $21.00.

The dominant Western paradigm on the Arab-Israeli conflict treats the Palestinian issue as a by-product of the conflict rather than its root cause. In the various contortions, convulsions, and explosions of the conflict over the last forty years, the so-called Palestinian problem has been viewed from this perspective, shared by both U.S. and Israeli analysts, as tangential to a solution. This is nowhere more clearly manifested than in the Camp David Accord.

The Palestinian-problem paradigm has allowed abstraction of the Palestinian problem from the context of the Arab-Israeli conflict, and it has allowed disaggregation of the multifaceted pathos and fragmented fate of the Palestinian people—so the West Bank, Gaza, the Palestine Liberation Organization (PLO), the Palestinian refugees, Jerusalem are viewed as discrete issues rather than as dimensions of a whole. This paradigm is posited on the assumption of the Palestinian issue as an object of the Arab-Israeli conflict rather than the subject of the conflict.

The two volumes under review here reflect the serious limitations of the paradigm. Don Peretz's monograph, *The West Bank*, is a disciplined effort to delineate systematically all of the dimensions "in the search for a West Bank solution." The study, originally undertaken by a research team and compiled and edited for this volume by Peretz, was contracted by the U.S. Defense Intelligence Agency. Its terms of reference were no doubt limited to the West Bank by that contract.

Peretz uses a diachronic—pre-post Israeli occupation—framework to organize the study. Part 1, "The West Bank: The Late Ottoman Period through 1967," divides the history of the area into three periods: the Ottoman era (chapter 1), the mandatory era (chapter 2), and the Jordanian era (chapter 3). In the introduction to this part, Peretz notes that for the period before 1950 it is "almost impossible to separate the life patterns of the West Bank inhabitants from Arab Jerusalem or the other regions of Arab Palestine." But with the creation of the State of Israel in 1948, the West Bank emerged as a problem—first for Jordan, then for Israel, which was apparently provoked into occupying the area in the Arab-Israeli war of 4-5 June 1967. But it was Jordan's "loss of the West Bank," not Israel's military occupation of it, that "resulted in a major escalation of the Arab refugee problem and the total transformation of the demographic, political and economic face of the West Bank," according to Peretz.

With the causes of the problem essentially established in part 1—that is, with the West Bank effectively delineated as a distinct problem in and of itself—part 2 addresses Israel's handling of the problem since its occupation of the area. The first five chapters in this section address the various dilemmas, successes, and failures of Israeli policy in the West Bank. The overall picture presented is one of Israel struggling with the conflicting demands of the Israeli political constituency (chapter 4), the dynamism of Jewish settlement policies (chapter 5), the efficiency of military administration (chapter 6), the declining fertility and improving welfare of a hostile population agitated by outside forces (chapter 7), and the outcomes of a colonialist economic policy (chapter 8)—struggling with all this to find a solution to the West Bank problem. Presented as obstacles to a solution rather than as subjects of the problem, Israeli policies in the West Bank are treated as effects inhibiting resolution rather than as causes requiring resolution.

Chapter 9 considers options that have been proposed for the future of the West Bank. These range from annexation by Israel to establishment of an independent Palestinian Arab state, including a number of intermediate proposals between these two extremes. In assessing these options, Peretz points out that none of them considers the role of Jerusalem, Gaza, or the Arab refugees; therefore, none of these options is likely to contribute substantively to a resolution of the Arab-Israeli conflict.

This monograph presents a systematic study of the West Bank problem within the framework of a particular paradigm, and it presents the options for resolution that logically derive from this approach. The conclusion, however, reveals Peretz's larger concern with resolution of the Arab-Israeli conflict rather than the West Bank problem, and he is rightly concerned with the potential dangers of a piecemeal approach. But he can only present this as a dilemma, confined as he is by the biases and distortions inherent in the paradigm.

Shaul Mishal's book, *The PLO under Arafat,* reflects another variation of the Palestinian problem paradigm. The PLO problem, this book's focus, takes the organization itself as the general problem and Arafat's leadership of it as the specific problem. The categories of analysis are essentially structured by this definition of the problem—the organization and its environment, the man and his ego. As Mishal explains in his preface,

> Arafat has hesitated to adopt a new and daring strategy that would maximize his chance to participate in a negotiated settlement. To gain a better understanding of Arafat's reluctance to take an innovative path, one must examine the nature of the PLO's internal and external political environment. Throughout this book I will elucidate Arafat's inability to assure control over developments in the two environments and his consequent fears that a daring strategy would not improve his chances of translating PLO diplomatic successes into territorial gains (pp. xiii-xiv).

Thus the study is a curious mixture of organizational, social, and psychological variables to explain what Mishal perceives as "Arafat's irrational behavior."

Chapter 1, "Dilemma of the Disinherited," examines the formation of the PLO, its goals and changing strategies—particularly those of armed struggle—up to the 1973 Arab-Israeli war. Chapter 2, "The Fragmented Basis of Palestinian Nationalism," examines the various factions that the PLO represents and the conflicting interests they encompass, especially in the post-1973 period when the debate over armed struggle versus political settlement intensified. While these two chapters explore the "internal constraints" and "institutional anxieties" that structure PLO goals and strategy, chapter 3, "Diplomacy in Chains," focuses on social and psychological constraints and anxieties "that make it almost impossible for the Fatah leadership publicly to accept the idea that Palestinian national goals will be fulfilled in less than the whole territory of Palestine"—even though they are privately pursued, as Mishal demonstrates.

Chapter 4, "In the Web of International Diplomacy," examines the PLO's response to five international peace proposals. While the PLO's public position on concessions to Israel—irrespective of its demonstrated flexibility with regard to the peace proposals—is essentially explained as a neurotic commitment arising out of internal tensions, fears, and anxieties, apparently Israel's intransigent position and heavy-handed policies vis-à-vis the Palestinians are also explained by the PLO's neurosis:

> However, such an assessment, whatever role it played in the PLO's diplomatic success, did not shake the Israeli government's belief that the organization remained faithful to its ultimate vision of a Palestinian state in place of Israel. Guided by such a belief, Israel took political and military measures in the occupied territories to reduce the chances of the establishment of a Palestinian state (p. 96).

Chapter 5, "The PLO and the West Bank: An Uneasy Alliance," and chapter 6, "PLO and West Bank: Bridge over Troubled Water," examine the scope and nature of PLO influence on the West Bank, suggesting in effect that PLO support there is more apparent than real and PLO motives more self-serving than West Bank-serving. Chapter 7, "Why Not a Daring Strategy?" essentially brings the text back to the question posed in the preface and summarizes the findings, concluding that the explanation of the PLO problem lies within the PLO and the Arab environment that spawned it. The sociology of knowledge has developed the metaphor "blaming the victim" to explain this kind of reasoning. It seems appropriate here.

TAREQ Y. ISMAEL
American University in Cairo
Egypt

RAMAZANI, R. K. *Revolutionary Iran: Challenge and Response in the Middle East.* Pp. xv, 311. Baltimore, MD: Johns Hopkins University Press, 1986. $27.50.

ZABIH, SEPEHR. *The Left in Contemporary Iran.* Pp. 239. Stanford, CA: Hoover Institute Press, 1986. $24.95.

In his short foreword to Ramazani's *Revolutionary Iran,* Stanley Hoffman has suggested the alternative title of "Coping with the Iranian Revolution." Indeed, the primary purpose, as well as the general frame of reference, of this book is to ascertain the political and diplomatic implication of the Islamic Revolution in Iran not only for Iran's immediate neighbors, discussed in parts 1 through 3, but for as distant an interested party as the United States, discussed in parts 4 and 5.

In part 1, Ramazani identifies the ideological posture of revolutionary Iran and measures its significance both in terms of the radical Shiite responses to its resurgent leadership and in view of the immediate threat facing such strategically sensitive areas as the Strait of Hormuz and Saudi Arabia. Part 2 focuses on the Persian Gulf as the political and military center of gravity that projects the Iran-Iraq war into its wider implications. The ramifications of the Iranian revolution and the Iran-Iraq war for the Eastern Mediterranean are considered in part 3, and in chapters focusing on Israel, Egypt, and Lebanon. Part 4 examines the dual implications of oil for both the revolution and the war. Finally, part 5 offers a few guidelines for U.S. policy toward this highly sensitive part of the world.

Two central issues constitute the main motifs of this study: the revolutionary and defiant nature of "the ideological crusade" established in Iran; and the ramifications of this crusade for the rest of the world. Ramazani's painstaking research and exhaustive arguments leave no doubt that there indeed exists what he calls "a sociopolitical explosion" in the region. But given the volatile and highly elusive nature of politics of the region, neither "the ideological crusade" is as deeply committed as Ramazani imagines nor are the global implications of this crusade as persistent and widespread as he argues.

The leaders of the Islamic Republic are much too serious students, if not masters, of *realpolitik* than to take Ramazani's formulation of Khomeini's ideas—that "the present international system must be transformed into the abode of humankind and, above all, the home of the 'oppressed masses of the people' "—too seriously. The Iran-contra scandal makes a total mockery of the rhetorics of the slogan "Neither East, nor West, only the Islamic Republic," which, alas, Ramazani takes too seriously.

The global implications of this highly volatile and essentially pragmatic phase of the revolution are grossly exaggerated in Ramazani's study. There are geopolitical, economic, and even racial dimensions to the Iran-Iraq war that cannot be totally reduced to the mere catch phrase of "Iraq's response to the perceived threat of contagion of the Khomeini brand of Islamic fundamentalism." During the late Shah of Iran's time, Iran and Iraq were ostensibly on the brink of war on more than one occasion, obviously without the presence of any Islamic fundamentalist factor. For years the late shah financed Mulla Mustafa Barezani's Kurdish rebellion against the Iraqis; Iraqis reciprocated with Kurds in Iran. Certainly, those phases of the Iran-Iraq war lacked any obvious religious implications. Given the deep-rooted delineation of geopolitical, economic, sectarian, and racial configurations of the Middle East, no particular event such as the Iranian revolution, even if it had a sustained ideological thrust, can effectively change the established political relations.

The significance of the Iranian revolution in its geopolitical context, of course, cannot be denied. But a frame of reference much wider than the immediate aftermath of the revolution is necessary to venture "guidelines for U.S. policy"—if indeed that is a tenable academic objective.

Sepehr Zabih's *Left in Contemporary Iran* is the latest study by a veteran student of radical movements in Iran. After presenting a short historical background and a discussion of the "Soviet connection," Zabih in four successive chapters examines the fate of the Tudeh Party in postrevolutionary Iran, the growth and organizational vicissitudes of the People's Mojahedin, the origin and revolutionary doctrines of the Fedayeen, and, finally, the militant tendencies of the League of Iranian Communists.

Like his previous studies of the Left, notably *The Communist Movement in Iran* (1966), Zabih's updated account is well documented and thorough. The essential part of the study deals with the postrevolutionary fate of the Left in Iran. The Tudeh Party is shown to have tried to resuscitate its past significance on the Iranian political scene by supporting the Islamic Republic against other leftist groups. Having utilized the Tudeh Party to suppress its opponents, the Islamic Republic did not spare this transitory ally and finally subjected its leaders to public humiliation and confession of treason.

The People's Mojahedin, too, were soon to become disillusioned with the Islamic Republic once they realized that, despite their contribution to the revolution, they were not to be beneficiaries of its power. The Islamic Republic successfully crushed its internal organization under the military command of Mussa Khiyabani and forced its leader, Masud Rajavi, into exile.

The veneer of Islamic appearance distinguished the People's Mojahedin from the more patently Marxist organization of the Fedayeen. Zabih traces the origin of this group back to its inception in 1963 by Bijan Jazani. Following their urban guerrilla activities during the 1960s, the Fedayeen became actively involved in the 1979 revolution but were equally denied any share of the power. In response to the Islamic Republic, the Fedayeen split into two factions: the majority supporting the Islamic Republic, and the minority opposing it. Further developments in the course of the Islamic Republic caused further factionalism and effective retardation of this group.

Zabih's conclusion is that the most important leftist organization that along with the army and the pro-Islamic Republic forces will shape the future political scene in Iran is that of the People's Mojahedin.

Zabih's volume is indispensable reading for those who follow the tumultuous and erratic saga of the Left in Iran.

HAMID DABASHI

University of Pennsylvania
Philadelphia

SOMMERVILLE, CAROLYN M. *Drought and Aid in the Sahel: A Decade of Development Cooperation.* Pp. xxiii, 306. Boulder, CO: Westview Press, 1986. $28.50.

This publication is a reproduction—it appears to be more or less an exact replica—of Sommerville's doctoral dissertation, and it is a good one. It is not clear from which university Sommerville earned her Ph.D.

This is not going to be a book review in the usual sense of the term. We are going to excerpt from the preface of this book in order to give the reader an idea of the main issues that the book deals with in regard to Third World regional efforts in addressing certain problems of drought and desertification, and the cooperation among the donor countries in responding to the problems of the African countries involved.

To complete the research for this study, Sommerville initially spent six months in the Sahel and Paris in 1980, returned to Paris for two weeks in 1982, with an additional trip there in 1985 to update materials following recurrence of drought in the Sahel in 1982-85.

Sommerville notes that intellectual themes

and research issues pertaining to "development and underdevelopment, forms of international cooperation, the structure of international organizations and the role of aid in development are the issues," and "in the Comité permanent Interetats de Lutte contre la Sécheresse dans le Sahel. . .—CILSS—and its parallel donor group, the Club du Sahel, all of these issues converge." The book is a case study of CILSS, its response to the drought of 1968-74 and attendant famine in West Africa—membership of CILSS is composed of the most affected countries: Burkina Faso, Cape Verde, Chad, Gambia, Guinea-Bissau, Mali, Mauritania, Niger, and Senegal—and its relevance as a model for international cooperation.

Despite some amelioration of the problems of these West African nations that has resulted from the cooperation between CILSS and the Club du Sahel, "the region is further from the goal of self-sustaining economic development" than it was a decade ago. Sommerville maintains that the unintended—and, shall we say, unanticipated—by-product of the aid from the donor countries for the recipient countries has been "increased dependency: food self-sufficiency has declined to the lowest levels in decades and the reliance on food aid and food imports has become an enduring feature of the political economy of the Sahel." Development aid has not always been judiciously used for agricultural improvement, and it has often been used for nonproductive activities. Indeed, Sommerville is of the opinion that the increase in aid—the overseas per capita development assistance to the Sahel has surpassed the levels given to Asian and sub-Saharan countries—has contributed to a growing debt burden facing the countries of the Sahel.

Per capita development assistance by the Club du Sahel to the nations of the Sahel may have surpassed the level of aid given to many countries, but the failure of aid efforts to ameliorate the problems faced by the West African countries cannot be attributed to too high a level of aid. Rather, the problem probably is that the level of aid given by the donor countries was not sufficiently high, and at the same time not sufficiently sophisticated, to meet the gravity of the drought and its attendant social and economic havoc.

SURINDER K. MEHTA

University of Massachusetts
Amherst

TUTINO, JOHN. *From Insurrection to Revolution in Mexico: Social Bases of Agrarian Violence, 1750-1940*. Pp. xx, 425. Princeton, NJ: Princeton University Press, 1987. $42.50.

From Insurrection to Revolution represents a nice blend of social science and historical analysis. The introduction provides a good, up-to-date summary of the literature dealing with the causes of rural rebellion and then proposes a new theoretical synthesis. In the rest of the book, Tutino analyzes the history of insurrection and revolution in rural Mexico from 1750 to 1940. Using a variety of case studies from different regions, he explains why some farmers, peasants, rural laborers, and craftsmen became violent insurgents or joined revolutionary movements while others stood on the sidelines or even opposed rebellions. In going from decade to decade, Tutino shows how sporadic peasant uprisings eventually spread throughout most of rural Mexico. The reader also learns that the focus of agrarian militancy shifted from one region to another. While some parts of rural Mexico, such as the Sierra Gorda, continued to display rural rebellions, other regions of popular uprisings in the nineteenth century were quiet even during the violent decade of civil war and peasant rebellion from 1911 to 1924 known as the Mexican Revolution and vice versa.

Tutino's theoretical synthesis combines previous approaches that emphasized subjective factors—for example, how rural lower classes perceive class relations and their notions of injustice—with a perspective focusing on changing economic and political structures in a more objective manner. He stresses the complex interplay between four variables—material conditions, autonomy, security, and mobility—in the process leading to peasant rebellion or passivity. For example, Tutino argues that rural people with little autonomy or mobility but a great deal of security are just as unlikely to rebel as those who have little economic security but a great deal of autonomy. Long-autonomous peasants are more likely to become militant in situations "where their security is undermined by the evident acts of powerholders—and where their loss of security is not compensated for by a new mobility." The crucial determinant is whether or not peasants attribute declining standards of living, famine, or the negative impact of rapid socioeconomic change to identifiable human agents.

The historical period analyzed in most detail is the era just before independence and the early part of the nineteenth century. Representing almost half of the book, this section includes original data collected by Tutino himself. The rest of the book relies more heavily on secondary sources to illustrate his central thesis. Consequently, some of the material on the revolutionary period does not sound as convincing. These minor flaws, however, do not detract from the utility of the overall analysis.

The theoretical model Tutino develops can stand by itself and is certainly applicable to countries other than Mexico. This model now needs to be further refined. For example, I would like to see a more precise definition of rural mobility. I am not always sure whether Tutino is referring to a greater choice of place of work or occupation—geographical mobility—or the opportunity to improve one's economic position and status—social mobility. Surely, intergenerational mobility—that is, perceived improvement in the life chances for one's children—should also be taken into consideration?

Notwithstanding my critical comments, Tutino's book is well written and provides an original theoretical synthesis, bridging the gap, in the field of rural studies, between what has been labeled as the moral economy approach, associated with Scott, and the so-called political economy approach of Popkin. I strongly recommend it as an analysis of the agrarian history of modern Mexico, with many new insights. Apart from its logical precision, this book raises a host of questions and suggestions for further research.

FRANS J. SCHRYER

University of Guelph
Ontario
Canada

EUROPE

BIALER, SEWERYN. *The Soviet Paradox: External Expansion, Internal Decline.* Pp. ix, 391. New York: Alfred A. Knopf, 1986. $18.95.

While the notion of a Soviet paradox is certainly not novel, it has yet to be so articulately and insightfully developed as it is in Seweryn Bialer's most recent work, *The Soviet Paradox.* As Bialer delineates it, the Soviet paradox lies in the coincidence of the height of Soviet military power and serious internal decline. The system that has successfully struggled for generations to create one of the world's two most powerful militaries is eroding at its core, just as it has obtained its greatest international power and prestige. The very system that built contemporary Soviet power is itself in dire need of repair. Furthermore, the utopian dreams that originally impelled the Soviet leaders to seek international changes have long faded. This is the dilemma now facing General Secretary Gorbachev and his associates. After carefully tracing the historical roots of the Soviet paradox, Bialer addresses two vital questions. First, to what extent is Gorbachev likely to resolve the Soviet paradox? Second, what future implications does the Soviet paradox hold for Soviet foreign policy in general, and Soviet-American relations in particular?

From Bialer's perspective the Soviet paradox is deeply rooted in Soviet history. The Stalinist model of economic and political centralization remains a central component of the contemporary Soviet system. It was the instability of Khrushchev's rule that prompted the extreme stability, even paralysis, of the Brezhnev years. It was the stability of the Brezhnev years that prevented the changes in the system necessary to avoid internal decay. The Soviet system has evolved, adjustments have been made, and modifications have been imposed, but the essential components of the system—centralized power and a command economy, for example—remain. If the rate of growth in the Soviet military-industrial complex is to be sustained, however, a number of basic changes will have to made; alterations and modifications will not suffice. As Bialer notes, the revitalization of the system requires a devolution of power. Yet the centralization of power has been the leadership's paramount goal since its inception.

The question, then, is, Are Gorbachev and his associates willing to enact the necessary changes and capable of doing so? Bialer's answer is no on both accounts. Gorbachev, like his mentor Yuri Andropov, seeks to improve the system through improved organization and planning, an infusion of new discipline, and the eradication of corruption. Moreover, even if Gorbachev was resolutely in favor of systemic change, the bureaucratic and other impediments to such change are considerable. Only the unlikely advent of an unchallengeable leader like Stalin would allow such change. At this point in his argument, Bialer admirably resists the all too prevalent sensationalist conclusion. A seriously flawed system is not a failed one and decline does not signal collapse. With serious adjustment and modification the Soviet Union can certainly muddle through, perhaps in a form much revitalized relative to the Brezhnev years.

For Bialer, the implications of the Soviet paradox for Soviet foreign policy are both numerous and profound. Briefly, in light of declining Soviet foreign policy resources, pragmatism and caution, already trademarks of Soviet policy, are likely to be reinforced. More novel are Bialer's recommendations for American policy. Principally, American decision makers must realize that America cannot alter Soviet internal developments or the sources of Soviet foreign policy. To attempt this is not only fruitless, but dangerous. In this sense, the assumptions behind America's détente policy of the 1970s and its harder line of the 1980s are equally flawed. Each, in its own way, sought to change Soviet behavior in a relatively short time period. Instead, Bialer advocates what he terms "competitive coexistence" or "managed rivalry." Specific Soviet policies that threaten American interests can, and should be, vigorously opposed. But, at the same time, this rivalry must be actively managed and contained to avoid direct confrontation and its incalculable risks.

For Bialer, competitive coexistence necessitates a revitalization of the disarmament process. It is this last conclusion that is perhaps most troubling. While the reader may find it intuitively compelling, it is not clearly derived from the preceding line of argument. Arms negotiations are but one means of managing a rivalry. Bialer's assertion that disarmament talks and continued Strategic Defense Initiative research are two irreconcilable options is similarly controversial. Whether further American Strategic Defense Initiative research might result in a deepened and more earnest Soviet interest in negotiation remains an open question. These, however, are but minor criticisms of an otherwise masterful work. While some of Bialer's conclusions are certainly subject to challenge, overall, Bialer provides one of the most comprehensive, articulate, and balanced analyses of the Soviet Union today. This book is a modern classic.

DANIEL R. KEMPTON

Northern Illinois University
De Kalb

PEPPER, SIMON and NICHOLAS ADAMS. *Firearms and Fortifications: Military Architecture and Siege Warfare in Sixteenth-*

Century Siena. Pp. xxiv, 245. Chicago: University of Chicago Press, 1986. $24.95.

The deceptive narrowness of its title notwithstanding, this book offers important new insights into the nature of early modern warfare. A case study with broad general implications, this work displays a tightness of focus as impressive as the skills that the authors, an architect and a historian, bring to bear. Using artistic and architectural evidence—the maps, many keyed to contemporary paintings, and drawings are particularly effective—as well as documentary sources, Pepper and Adams evaluate the response of a pivotal Italian polity to gunpowder, revising traditional interpretations in the process. They show how siege operations were supported—and restricted—by a constant little war of patrols, raids, and blockade-running. Their explanation of the development of the angle bastion is new and sophisticated, properly deemphasizing the role of individual creative genius in fortress design; fortifications were, with rare and minor exceptions, collective undertakings shaped by broad economic, political, social, and technological considerations, not works of art conceived and executed by a single draftsman. The operationally realistic analysis of early sixteenth-century artillery is in refreshing contrast to the general tendency to exaggerate its effectiveness, based on the initial impact of Charles VIII's siege train on Italy in 1494.

The narrative is highly readable. Pepper and Adams handle logistical and motivational factors skillfully. Their command of primary sources is impressive and their knowledge of the secondary literature is encyclopedic. Their use of physical and graphic evidence is particularly noteworthy and the maps, drawings, and photographs are beautifully integrated with the text. The book is attractively laid out and well produced.

Firearms and Fortifications constitutes an important revision of accepted interpretations of the role of the fortress and military engineering in the early modern world. It significantly advances and refines our knowledge of the operational use of gunpowder weapons and of the conduct of positional warfare in the first half of the sixteenth century and may be regarded as definitive in this respect. Its theses, many of them controversial, are convincingly argued; they bear not only on military history and the history of technology, but on art history, the history of architecture, and the history of ideas. Political scientists engaged in conflict studies and with an interest in broad historical comparisons will find this book of great value; it is highly recommended.

JOHN F. GUILMARTIN, Jr.
United States Naval War College
Newport
Rhode Island

Ohio State University
Columbus

RITCHIE, ROBERT C. *Captain Kidd and the War against the Pirates.* Pp. vii, 306. Cambridge, MA: Harvard University Press, 1986. $20.00.

This book may be regarded as the definitive work on the life of Captain William Kidd. Robert C. Ritchie's research among the relevant sources in London in the Public Record Office and India Office records on Kidd's career has been very thorough indeed. An authority on American colonial history and especially New York's colonial history, Ritchie gives the reader a vivid picture of an able sea captain's life in the late seventeenth century. Kidd, often a pirate, had the bad luck to be tried and hanged for piracy on his return to London in the spring of 1701, despite his hopes of pardon through the intercession of friends in high places. Mere chance may have had the most to do with his later notoriety. It was not entirely his fault that his and Morgan's were the only names that stood out in the minds of later generations of English school boys.

Ritchie's picture of piracy at the close of the seventeenth and the beginning of the eighteenth century is not quite so clear. The uninitiated reader should prepare for it by refreshing his or her mind on British history in the age of William and Mary. Moreover, all readers would have likewise benefited by more briefing about the role of piracy in the age of mercantilism. Its part in smuggling goods into Britain from French and Irish ports is hardly mentioned. Piracy's background in the Atlantic and Caribbean is far more thoroughly dealt with than its background in the Indian Ocean. Piracy's role in circumventing the Navigation Acts by facilitating the direct import into the Americas of slaves from Africa, and fine cotton piece goods and other oriental products from India, receives far more attention than does buccaneering beyond the Cape of Good Hope. Captain Kidd's visits to the pirate nests in the islands close to Madagascar are well described, but the broader picture receives little attention. No mention is made of the coarse—then called "gruff"—India cotton piece goods, most of which had to go all the way to London in East India Company cargoes and then back down the coast of West Africa in Royal African Company cargoes to purchase slaves from African tribal chieftains, by whom such gruff goods were in great demand. Piracy in the West was already meeting a piracy in the East; this was soon to result in the foundation of the British East India Company's Bombay Marine, the first modern Indian navy.

HOLDEN FURBER
University of Pennsylvania
Philadelphia

SEIGEL, JERROLD. *Bohemian Paris: Culture, Politics, and the Boundaries of Bourgeois Life, 1830-1930.* Pp. ix, 453. New York: Viking Penguin, 1986. $35.00.

Seigel argues that Bohemian Paris—the Bohemia of Bohemias—was not the sworn enemy of middle-class life but, indeed, was the child of middle-class prosperity, democracy, and individuality. As middle-class society took form in nineteenth-century Europe and France, so too did Bohemian Paris take form.

According to Seigel, nineteenth-century Bohemian Paris was composed in majority of rebellious youth from middle-class France, and Bohemia's spokesmen in the main knew that they had their origins in the middle class. They not only did not stand for permanent opposition to the middle class but sought reconciliation with and recognition from bourgeois society.

Additionally, in Seigel's opinion, Bohemia's residents were not in majority talented painters or artists willing to be permanently impoverished and dedicated to spending a lifetime at the margins of society. On the contrary, many of its residents—Courbet and Picasso, to name the most famous—chose Bohemia as the workshop where they prepared themselves for bourgeois fame and wealth.

Bohemia's representatives did not live by Vigny's self-indulgent accusation that set the suffering of every sensitive artist at the steps of an insensitive middle-class society. Nor did they in significant numbers take their place at the barricades in the revolutions of 1848 and 1871.

For Seigel the essence of Parisian Bohemia was a great experiment with individual freedom, which itself was made possible by emerging industrial capitalism, liberal democracy, and middle-class society. In Bohemia the rebellious youth of France and Europe found a place where they could go and do what was not possible at home. There the "boundaries of bourgeois life" were tested as nowhere else.

Seigel, whose work develops around a succession of critical biographies, presents the shifting nature of Bohemia and the changing nature of its experimentation in reference to writers, poets, and artists, as well as critics, art dealers, and cabaret owners. There we meet such notables as Baudelaire, Courbet, and Verlaine, such lesser figures as Murger, Goudeau, and Valles, and such critics as the brothers Goncourt, Barrès, and Zola.

For Seigel, World War I was a turning point in the history of Bohemia. The war severely diminished the freedom of Bohemia by irreversibly mobilizing the hearts and minds. Opinion prevailed, ideology triumphed, and spontaneity was lost. The distance between art and life, so essential to classic Bohemia, was lost.

By the middle 1920s Dada, itself an ideological and didactic experimentalism, quickly gave way to surrealism, which under the leadership of Breton, surrendered to communism. More directly connected to the middle class and scientific ideas than its artistic predecessors, the surrealists wrote manifestos on behalf of a committed art.

Bourgeois civilization had matured. The age of innocence was over. Each decade after the 1920s found artists taking their opinions about their art and everything else ever more seriously. Mass, commercial, national, academic, and scientific cultures left room for the spontaneity of old Bohemia. Rebellion became standard, experimentation consciously contrived, and shocking the bourgeoisie became the oldest game in the whole world.

According to Seigel, the remnants of Bohemia and its imitators are now scattered throughout the West. In one sense Bohemia has won out: it has a place in every middle-class soul and in the corner of almost every shopping mall. In another sense, like traditional folk cultures, Bohemias everywhere have lost their autonomy and no longer are the preserves of a unique group.

Seigel is a sure and thoughtful guide to nineteenth-century Parisian Bohemia. A social-cultural historian hypothetically might have preferred a work dedicated to creating a collective material portrait of Bohemia's residents—their numbers, origins, housing, jobs, male-female mix, and the intersection of institutions—and a more careful analysis of the interplay of emerging national, democratic, industrial, and commercial as well as academic and scientific cultures. Yet these preferences, which may well be methodologically impossible given the shifting nature of Bohemia and the type of evidence needed, are not meant to distract from the worth of Seigel's important volume. His book belongs not only in the library of the intellectual and cultural historian but in the library of everyone who is seriously interested in the matter of individual self-definition in modern society.

JOSEPH A. AMATO

Southwest State University
Marshall
Minnesota

WILSON, TREVOR. *The Myriad Faces of War: Britain and the Great War, 1914-1918*. Pp. xvi, 864. Cambridge: Polity Press; New York: Basil Blackwell, 1986. $24.95.

Trevor Wilson, professor of history at the University of Adelaide, Australia, has spent the last ten years writing the first reevaluation of Britain and the Great War in as many years. The result is a staggering compilation of more than 800 pages in small print. Wilson's major thesis is that World War I for Britain in many ways was a good war, much as World War II was to be. Britain maintained the balance of power in Europe, gained an enhanced position in the Middle East, eliminated the German navy as a potential threat to its vital maritime lifeline, and at least for the time being retarded the dissolution of its empire as well as the global expansion of the United States. All this, Wilson argues, was accomplished without abandoning liberal precepts and the parliamentary system at home. Indeed, he suggests that the government of Herbert Asquith had no alternative but to enter the war in 1914; failure to do so would have resulted in the extermination of liberties in Europe under a German military autocracy.

The real value of the book lies in its judicious combination of domestic and foreign policies, in its treatment of the real war in France as well as the home front in Britain. Thus we are given not only the standard accounts of slaughter in the trenches, but also the social, economic, and demographic effects of the war upon English village life—Great Leighs, Essex—upon labor especially in the coal mines, and upon the growing female work force in the transportation and munitions sectors. Nor does Wilson overlook the propaganda war conducted by the Ministry of Information—the forerunner of Joseph Goebbels's less euphemistic Ministry of Propaganda.

Unfortunately, the book is deprived of much of its value to the professional scholar because of a dearth of footnotes and the lack of a decent bibliography. The Imperial War Museum, for example, is the only archive cited in the bibliography at the start of the book—along with several collections of private papers. The individual chapters are chock-full of direct citations and provocative hypotheses that remain largely undocumented. The tremendous outpouring of literature on World War I in the last two decades, not only on Britain but especially on France and Germany, for instance, is completely ignored, as far as one can tell. This is unfortunate, for the book is well written, cogently argued, and a treasure trove of information. Above all, it offers a far better picture of modern war than John Keegan's highly celebrated *Face of Battle* (1976).

HOLGER H. HERWIG

Vanderbilt University
Nashville
Tennessee

UNITED STATES

BERINGER, RICHARD E., HERMAN HATTAWAY, ARCHER JONES, and WILLIAM N. STILL, Jr. *Why the South Lost the Civil War*. Athens: University of Georgia Press, 1986. $29.95.

Question framing is elemental in scholarship; it predicts outcome via design. The authors of this fascinating book have chosen to frame their title question within the troublesome tradition of American Civil War historiography: that is, their main quest is fictional and metaphysical. Could the Confederacy have won the war? If not, was defeat inevitable? The answers are yes and no. "With greater effort southerners could have won their independence," they write, "and like most historians we have an aversion to the claim that great events have inevitable outcomes.

Much of the volume is devoted to elimination of conventional explanations for the Confederacy's demise. The South was not defeated by the North's overwhelming population and military superiority. Southerners enjoyed the enormous advantage of fighting a defensive war. Nor were Southern military tactics faulty. Here Beringer and his colleagues take the late T. Harry Williams to task, rehashing and reapplying Jomini's and Clausewitz's doctrines, which Williams misunderstood. They also eliminate the Union blockade of Southern coastlines as a significant cause of Confederate defeat. Likewise, they attack the died-of-state-rights school of Southern frustration. State rights—in the form of state relief and other services to the yeoman majority—actually strengthened Confederate resolve.

The authors' positive argument is controversial: the failure of Southern independence was a failure of will. Will consists in three connected parts. First, Southern nationalism was weak, its culture and icons too intermixed with American nationalism. Second, Southern evangelical religion was fatalistic. Confederates were foreordained to see God on the Yankees' side when the war went bad, particularly in the fall of 1864. Third, a significant number of white Southerners felt guilt over slavery, and once Lincoln changed Union war policy to abolition as well as union, the Confederates' own scruples about the peculiar institution undermined their resolve to fight on. That the Union's superior force might have caused failure of will the authors do not acknowledge.

Why the South Lost the Civil War is more compendious than seminal. Its arguments, negative and positive, may nearly all be discovered on the endless shelves of Civil

War literature. The book's virtues consist in clarity of summary and in bravery of argument. There is virtue, too, in the ambitious scope of the volume, combining generations of scholarship both serious and silly. This may be the one book to read about the Civil War. The shame is the persistence of distracting fictional and metaphysical question framing.

JACK TEMPLE KIRBY

Miami University
Oxford
Ohio

BYERLY, VICTORIA. *Hard Times Cotton Mill Girls: Personal Histories of Womanhood and Poverty in the South.* Pp. x, 223. Ithaca, NY: ILR Press, 1986. $26.00. Paperbound, $9.95.

WOOD, PHILLIP J. *Southern Capitalism: The Political Economy of North Carolina, 1880-1980.* Pp. xiii, 272. Durham, NC: Duke University Press, 1986. $37.50. Paperbound, $12.95.

These two books are both about North Carolina and both have something to say about that state's economic development. *Hard Times* is an oral history collection about women who work or worked in the mill towns of North Carolina. Old and young, black and white, the women tell stories of rich detail about their lives. Byerly introduces groups of the well-edited recollections with short essays that make sense of them in relation to the history of the textile industry and the women's part in that history. The common themes of their lives—poverty, family discord, male oppression, and hard work—give evidence of the harsh economics of the textile industry. In a graceful introduction to Byerly and the women, historian Cletus Daniel writes that this collection helps us to understand in "irreducible form" the "unique sisterhood of hardship and struggle forged in textile mill villages of the Carolina Piedmont by the intersection of class and gender within the industrial continuum of the New South." More than that the reader will cheer for these women's gritty, dignified survival.

Phillip Wood's book is also about class and hardship in North Carolina, but his book is a complicated one of theory and argument wrapped around some very traditional history. Wood argues that neither traditional capitalistic development models nor a group of historians who study economic development in the late nineteenth-century South whom he calls the "new social historians" explains why the explosive accumulation of capital in North Carolina resulted in a hundred years of an "above-average rate of exploitation" of its working-class people. In theoretical sections he confidently contends that only a "framework of Marxian political economy" explains North Carolina's economic history and subsequently its social and political structures. Thus, throughout North Carolina's recent history, Wood discovers a capitalistic class leadership that relentlessly seeks surplus values that exceed the national averages by large margins, and these surplus values are extracted from ordinary workers in the form of below-average national living conditions, hard work, and, above all, low wages. The book makes no pretense to a comprehensive narrative. Rather, the first major part of the book concentrates, as a sort of case example, on the growth of the textile industry from the late nineteenth century to 1939. Following that, one fifty-page chapter explores the political relationship of the "state" to the developing economy from the Civil War through the New Deal years. Another long chapter argues that only Marxist theory makes sense of "state" economic development policy and the pattern of postwar industrialization. Oddly, strung throughout these chapters is a history of labor relations in the modern textile industry.

Wood's bold revisionism and the sweep of his argument make this book important. He spares no one. His disagreement ranges from V. O. Key to C. Vann Woodward to a bundle of America's brightest young historians. He has not entered any archives, but he has read widely and well and he floats stunning insights that stand independent of acceptance of his Marxist theory. As a theoretical work his book is short and simple; as a history narrative it is episodic, incomplete, and often wrong. But the book works; it provokes thought and involvement, and its brilliant juxtaposition of events, facts, and topics make it a must read for scholars of many persuasions. Just be prepared to argue and quibble. And after that read Byerly's collection for some solid satisfaction.

JAMES A. HODGES

College of Wooster
Ohio

CRABB, CECIL, V., Jr. and KEVIN V. MULCAHY. *Presidents and Foreign Policy Making: From FDR to Reagan*. Pp. xiv, 359. Baton Rouge: Louisiana State University Press, 1986. No price.

No American political office galvanizes more public or scholarly attention than the presidency. The unique concentration and scope of institutional power, modified by personality, style, capacity, and knowledge, make the wide and considerable interest in the presidency easy to understand. This timely study examines a paramount area of responsibility of the presidency, the conduct of foreign affairs, and thus deals with the president in his most powerful role. For as Crabb and Mulcahy make indubitably clear, in the area of foreign relations, the president, as granted by constitution, statute, and custom, comes closest to realizing a near monopoly of power.

This observation is the premise, however, of the delicate irony that forms the thesis of this book. While the president may be the virtual master of foreign policymaking, nevertheless, it is contended convincingly, policymaking in the executive branch over the past forty years has been marred by "conflict, fragmentation, and incoherence." Not surprisingly, then, recent history shows, as is clearly the case today, a lack of public consensus and a fractious questioning of the appropriate goals and objectives of American foreign policy. And finally, given the cogent picture of intra-executive hubris and rivalry, combined with a resurgent congressional interest in foreign policymaking, Crabb and Mulcahy maintain that there is an increased likelihood that defective directives will continue to flow from what arguably is an inherently flawed decisional process.

Recent developments, it is explained, have compounded the problem of rational decision making. Like the proverbial situation of a bevy of cooks in a small kitchen, the once relatively simple decisional agenda has been replaced by a variegated, complex, and unsuccessful diplomatic menu. The new critical dimensions of U.S. policy, in the form of military, technological, economic, and trade concerns, all heightened by superpower ideological competition, make for a greater number of vested interests as well as actual and potential players, in the most dramatic, high-stake, political game in town. But how the game is played does not seem to be affected by the particular issues. What is more, and this is the critical thrust of the book, irrespective of the administration in power and of party control in Congress, no matter how the president has organized his foreign policy-making apparatus, the results have been disappointing. Suggestions for future changes do not encourage great optimism.

Six informative case studies beginning with Roosevelt, who tended to act as his own secretary of state, and continuing through Reagan, who presented even before the Iran arms imbroglio an intriguing anarchical model of decision making, illustrate and undergird the main argument. The Ford term, perhaps for its briefness and lack of distinction, receives little attention. Less explicably, the Carter administration, with the exception of the celebrated Brzezinski-Vance rift, is essentially overlooked. Nonetheless, the material gathered, most of it from the broad swath of presidential scholarship, prudently poses the disconcerting Tocquevilleian conundrum, that is, whether democracy, which is far closer to justice than is the modern face of autocracy, can ever learn to manage as effectively as its competitor its life-sustaining and threatening external affairs.

PAUL L. ROSEN

Carleton University
Ottawa
Canada

EPSTEIN, LEON D. *Political Parties in the American Mold.* Pp. xiv, 440. Madison: University of Wisconsin Press, 1986. $27.50.

This book presents a well-reasoned and solidly documented statement on the nature and condition of American parties today and the forces that have molded their structures and operations. It contains a thorough discussion of the literature on political parties. Major authors and themes in that literature are discussed in historical perspective. And while the emphasis is on American parties, comparisons are made to other parties and party systems. The material goes beyond the traditional party literature in that works from other speciality fields of American politics are discussed if the institutions or processes they examine have affected, are affecting, or might affect American party development.

Epstein believes that American parties are not declining or dying. Instead he views them as reacting and adapting to the limits placed on them by constitutional, historical, and cultural forces. The adaptation process has been a continuous one and has followed an incremental pattern. By examining party development over time, he believes that we can come to a better understanding of how their adaptability has and will keep the parties alive.

The forces affecting the parties include public hostility toward both the machine-style party organizations of the past and the stronger party structures in the European style. These sentiments make changes such as those advocated by supporters of the responsible-party model unlikely to occur.

The book places greatest emphasis on the party as an organization. Separate chapters examine the national organization, state and local structures, the party as a public utility—that is, a regulated semipublic organization—and private and public funding of campaigns. The various reforms that have affected the party—for example, the direct primary, the Australian ballot, internal Democratic Party rule changes, federal court decisions—are examined in terms of both their causes and their effects.

The party in government is examined in two chapters on the presidency and the Congress. Here the historical development of the party system is considered.

The party in the electorate plays a minor role in the discussion. In the one chapter devoted to this topic, Epstein argues that the direct primary institutionalized the Republican and Democratic party labels and that candidate-centered campaigns have limited party loyalty. He contends that the labels still serve as cues for the electorate although to a lesser degree than in the past.

ANNE PERMALOFF

Auburn University
Montgomery
Alabama

GALLUP, GEORGE, Jr. and JIM CASTELLI. *The American Catholic People: Their Beliefs, Practices and Values.* Pp. ix, 206. Garden City, NY: Doubleday, 1987. $15.95.

WEIGEL, GEORGE. *Tranquilitas Ordinis: The Present Failure and Future Promise of American Catholic Thought on War and Peace.* Pp. xiii, 489. New York: Oxford University Press, 1987. $27.50.

Gallup and Castelli distill two decades' worth of Gallup poll data to forge a coherent social profile of America's Roman Catholics, 28 percent of the population. They contend—against journalistic stereotypes—that American Catholics are in the middle of a religious revival.

As the research of the National Opinion Research Center's Andrew Greeley has shown for decades, American Catholics are a highly mobile, well-educated, economically advancing segment of America. Yet, in several of their religious practices and elements of their worldview, Catholics represent a unique American segment. The Catholic worldview—so the statistics show—is more intellectual, pragmatic, and communal than Protestant America's. Moreover, American Catholics show "a remarkable degree of tolerance toward racial and ethnic minorities, followers of other religions and non-traditional sexual life-styles."

Catholics are reluctant to evangelize others but, today, "American Catholics no longer worry about being accepted—they worry about how to lead." Three notable sets of data demonstrate (1) basic support for the bishops' economic pastoral by Catholic rank and file; (2) basic support for the bishops' peace pastoral—Catholics are more strongly against defense spending and for peace education than others in America; and (3) despite a general pro-life stance, American Catholics exhibit strong support for women's rights.

Gallup and Castelli's portrait could, profitably, be contrasted with James Davison Hunter's sociological profile, *American Evangelicalism*, and with Wade Clark Roof's research on mainline Protestantism. Gallup and Castelli think American Catholics have made America a more racially and religiously tolerant society. "Stability and growth—not decline—are the earmarks of Catholic religious life today."

Weigel's *Tranquilitas Ordinis* devises an ideal-type analysis of a classic Roman Catholic heritage of a theory of ordered society, a doctrine of peace and just war that combines Augustinian realism with moral power. Along the way, Weigel constructs a historiographic straw man that, he claims, interprets pre-Vatican II Catholic support for American foreign policy as unthinking jingoism, acceptance by insecure immigrants. One is hard put to find a serious historian of American Catholicism—not Jay Dolan, James Hennesey, David O'Brien, or Gerald Fogarty—who would subscribe to this straw man. Thereupon Weigel fabricates a golden age of American Catholic realism—based on nursing a few citations from bishops' pastorals during World War II—which Weigel extols as an independent theoretical contribution to thought on internationalism and peace.

Everything in official American Catholicism since Vatican II—especially the American episcopacy's position on nuclear weapons and its policy in Central America—is, then, seen as the abandonment of the received heritage. The culprits are mindless neo-isolationism, anti-anti-communism, and superficial abandonment of the moral worthiness of the American experiment abroad in the American left that infects the bishops' thought. In Weigel, it is, at times, hard to distinguish criticism of specific American policies from anti-Americanism.

The book glides over the failures during the so-called classic period actually to apply the tradition to violations of *jus in bello*, for example, saturation bombing during World War II. Those who know Daniel Berrigan's strong and very public repudiation of Ernesto Cardinal because of the latter's support for violent revolution will be puzzled by the assertion on page 246 that Berrigan refused to dismiss violence in the Third World.

Weigel raises some valid analytic points—such as the easy conflation of pacifism and just war in the bishops' pastoral—but his polemic reading of post-Vatican II mainline American Catholic thought on peace will convince only those who can assume that Cardinal John Krol is a victim of Weigel's stipulated anti-anti-communism. More likely, Krol, like Tip O'Neill, is more apt to believe eyewitness church friends about El Salvador than administration spokespersons. A pity, in fact, that the sharp polemic detracts from what could encourage a serious analytic debate about the bishops' pastoral on peace.

JOHN A. COLEMAN

Graduate Theological Union
Berkeley
California

HANDLER, JOEL F. *The Conditions of Discretion: Autonomy, Community, Bureaucracy*. Pp. xi, 327. New York: Russell Sage, 1986. No price.

Few studies have probed or fused issues of law and public policy with such candor, sincerity, integrative insights, and attention to both theory and empirical illustrations as has Joel F. Handler in this volume of incisive but constructive reportage, critique, and prescription.

The task of distilling and analyzing legal, administrative, political, and philosophical dimensions of governance in relation to discretion could easily have yielded to pedantic or parochial urges. To his credit, Handler has probed a complex, amorphous concept's roles in society with originality, balance, and commendable aversions to formalism and pretentiousness. Part of the reason why, in his view, our nation's present approach to conceptualizing and administering justice "is doomed to failure" is that it places too much emphasis on formal institutional processes and practices, such as due process. He believes that the system of rights and procedural remedies developed over the last several decades does not work for the poor or for many with incomes above poverty definitions.

Handler's alternative to due process and the adversary confrontations it generates is not as lucidly set out as are his critiques of the status quo. Procedural due process is defective at present, he maintains, because it "sharpens rather than mutes adversarial relationships; it truncates and polarizes rather than heals; it is cross-sectional rather than continuous; it creates winners and losers rather than partners." The cooperative system he advocates to supplant procedural formalism is one that must countenance uncertainty and indeterminacy but is allegedly capable of achieving healing, continuity, and partnership through flexible and responsive utilization of discretion.

Chapter 4's forty pages provide an empirical analysis of the Madison, Wisconsin, School District's special education program as a potential model of cooperative decision making. By conceiving of parents as part of the solution rather than the problem, by positing acceptability of experimental decisions that can be accommodated and renegotiated, and by viewing conflict as a necessary component of communication for the reaching of understanding, the Madison approach, it is claimed, manifests theoretical coherence and operational practicality.

Handler recognizes that the Madison program's data also indicate "certain problems or weaknesses that are inevitable or inherent in this system." For example, the program reinforces parental passiveness. Routinization may set in, solidifying bureaucratic procedures and developing communitywide co-optation. "Trust breeds complacency" and can impair the quest for new techniques to maintain communicative conflicts and incentives for participation.

Handler's candor combines with the depth and sensitivity of his scholarship to make reading his book an engrossing and challenging venture for lawyers, philosophers, and social scientists at all career stages. It is an additional plus that his "worthy vision" of decisional structures that support participation, community, and individuality does not wither with the charge that it harbors utopianism. His skills are equally bountiful when it comes to propounding and defending his theses.

Notwithstanding the wide-ranging references to and invocations of scholars who have dealt with his subject, from Richard Abel to Robert Zupkis, it was a source of disappointment that salient works and thoughts by Michael Barkun, Jacobus ten Broek, and Kenneth Culp Davis were not among them. Consideration of Davis's contrasting position on the propriety of discretion, of ten Broek's compelling views on welfare law reforms, and of Barkun's insightful analyses of millenarian movements would have leavened Handler's discussions. Their absence, however, cannot be said to dent the creativity, humaneness, or pertinence of Handler's prescription for cooperative public action.

VICTOR G. ROSENBLUM

Northwestern University
Chicago
Illinois

KIRBY, JACK TEMPLE. *Rural Worlds Lost: The American South 1920-1960*. Pp. xix, 390. Baton Rouge: Louisiana State University Press, 1987. No price.

Thoroughly researched, this book tells in considerable detail the story of Southern agriculture and related industries during the decades 1920-60, with many chronological glances at conditions before and after these dates. It deals with the South as a region, with particular states within the region, and with counties within certain states. It also contains much material about individuals and families, by name. All this enriches the account but obliges the reader to guard against confusing the particular with the general.

Southern agriculture is and always has been varied, and much of the Southern terrain is suited to crop diversification. Cotton, always associated with the South, is also grown on the outside; indeed, in 1947 it was California's "most profitable crop." Corn, rice, sugar, tobacco, peanuts, soybeans, citrus and noncitrus fruits, and a variety of vegetables are produced in the South. The region is known also for dairying and livestock, for timber and timber products, for coal, and for numerous manufactures.

After the Civil War, most former slaves became sharecroppers and many landless whites became sharecroppers or tenants. The thirty years embracing World War I, the Depression, the New Deal, and World War II saw the Southern economy transformed. Gasoline-powered machinery reduced labor costs and caused large numbers of blacks and whites to move to Southern towns and cities or to big cities of the North. During the fifty years 1910-60, Southern migrants totaled over 9 million, slightly more than half of them being whites. As to the overall effect of the New Deal on the South, Kirby says that in

> predominantly white nonplantation areas, . . . the programs were inadequate as relief but positive and beneficial in the short run. In predominantly black plantation areas, . . . the programs rescued and enriched planter-landlords and inflicted frustration and suffering on the already poor and landless (p. 56).

While this book is mainly about Southern agriculture, it is also about the human beings who provided the labor for it all. It tells of malaria, hookworm, and pellagra, of doctors, midwives, and abortion, and of shabby, unsanitary habitations. It may be that only in the case of wives whose husbands served in World War II was there something approaching equity on a large scale. The wife received $50 a month and an extra allowance for each child. This may have been more cash on a regular basis than the family had ever received. One wonders whether there was much nostalgia for the Southern "rural worlds lost."

JENNINGS B. SANDERS

Kensington
Maryland

McCANN, MICHAEL. *Taking Reform Seriously: Perspectives on Public Interest Liberalism*. Pp. 345. Ithaca, NY: Cornell University Press, 1986. $29.95.

Michael McCann's *Taking Reform Seriously* is an original contribution to the burgeoning literature on interest groups. Rather than examining the origins or political strategies of these groups in great detail, it focuses on the political philosophy and dilemmas of the public interest groups and the activists who have organized them. Although McCann offers a basically sympathetic portrayal, he is concerned with illustrating the difficulties reform movements have in effecting significant political change. As such, his book is a cogent critique of the often self-imposed limitations on these groups. While not offering any new theoretical understanding of these movements, it does provide a different perspective for examining their behavior.

McCann argues that public interest groups in one sense see themselves as part of a pluralist democratic ideal that has been unfulfilled in practice due to the dominance of business and corporate interests. One strategy for asserting the public interest is, together with the mobilization of citizen resources, to promulgate a "doctrine of public rights" involving legal representation in government decision making in both the legislative and bureaucratic arenas.

McCann suggests close parallels between the philosophy of public interest liberalism and earlier reform movements. The suspicion of corporate power, moralism, middle-class status of the activists, and antiviolence toward governmental authority all reflect earlier movements, while the hostility in many cases toward growth, materialism, and the American world role are a break with the past. In my view, the parallels are overdrawn; the ideology of public interest liberalism fits much more closely with a New Class critique of American society. Also in that regard, McCann fails to deal adequately with the obvious complaint that these groups are really only involved in promoting their own variously defined group or class interests.

By far the most interesting aspect of this work is McCann's examination of why public interest liberalism's successes have been limited and even reversed. The most damning critique is his conclusion—I believe a correct one—that public interest liberalism has not developed a coherent overarching political philosophy. In particular, the consumerist ethic offers little in the way of solutions to the problems of economic growth and productivity, which these groups have normally taken for granted. And in the spirit of Michels, McCann shows that these groups, rather than live up to their participatory ideal, have been dominated within their organizations by a relatively small group of elites.

Last, McCann makes a strong case for the view that the highly conflictual orientation of these groups contributed to the "development of a structureless, polymorphous spongelike state. . . . lacking a viable institutional nervous system for converting . . . demands into consistent programs of publicity sanctioned goals and policies." This problem has been exacerbated by their suspicion of more institutionalized channels of expression such as the political parties. Thus these groups have been hoisted with their own petard and have contributed to a decline in the legitimacy of the liberal state while proclaiming the desire to enhance that very legitimacy.

Though slightly flawed, *Taking Reform Seriously* is an eminently readable, worthwhile contribution to the study of reform movements and their impact upon the policy process.

EUEL ELLIOTT

Virginia Polytechnic Institute
and State University
Blacksburg

MEAD, WALTER RUSSELL. *Mortal Splendor: The American Empire in Transition.* Pp. xii, 381. Boston: Houghton Mifflin, 1987. $19.95.

"The decline . . . of the American Empire is the basic political fact of the present period in world history." Mead's book analyzes the causes of this decline and asks what can be done about it.

The opening chapters describe the rise of the empire after World War II and the golden era of the 1960s when the liberal reforms begun by the New Deal made enormous progress. The basic institutions of the welfare state were established, and liberal ideas of equality and justice for all became the accepted goals of American society. All this was made possible by economic prosperity, and the national government, armed with tools of Keynesian economic theory, had mastered the techniques of economic management—prosperity was here to stay, profits and wages both rose, and class war was replaced by negotiation between employees and unions. Liberal capitalism appeared to be moving toward the successful democratic socialism of Sweden. The Third World was catching up rapidly; Russia would eventually join the club—it was only necessary to contain it during its temporary period of insanity. The world would soon become one great liberal society.

By 1980 this bright dream had faded away. Mead gives a superficial survey of what went wrong and then asks, "What can we do to be saved?"

Walter Russell Mead is a 35-year-old honors graduate of Groton and Yale, a political economist, editor, journalist, and teacher. He is a passionate advocate of the welfare state, democratic socialism, and the idea of progress. The builders of the American liberal empire were his aristocratic, Anglophile predecessors at Groton—Franklin Roosevelt, Dean Acheson, Sumner Welles, Averell Harriman, and the like. But this old-boy elite has lost its power and a new style of leadership must be found in the Democratic Party and the American Populist tradition. We "do not know how many of the lost sheep of liberalism can be reclaimed for the flock, but we can hope that a regenerate liberalism can say, with a much greater shepherd, 'My sheep know my voice.' " These new leaders will institute national economic planning—"new banking legislation will be the centerpiece of the New Liberalism"—the Third World will be set on the path of economic development, and the federal government, given vastly increased powers, will be streamlined for decisive action.

Mead's prescriptions for the future are vague, often contradictory, and always terrible naive. His is not a profound analysis of our ills—it reads like a collection of daily columns by a bright young journalist.

RICHARD SCHLATTER

Rutgers University
New Brunswick
New Jersey

MINK, GWENDOLYN. *Old Labor and New Immigrants in American Political Development: Union, Party, and State, 1875-1920.* Pp. 301. Ithaca, NY: Cornell University Press, 1986. $29.95.

In *Old Labor and New Immigrants in American Political Development*, Gwendolyn Mink asks a series of provocative new questions about the character and organization of the American labor movement during the late nineteenth and early twentieth centuries. Instead of trying to explain why American labor failed to develop along European lines, with an inherent class consciousness and a radical social program, Mink outlines how American racial and social conditions shaped a more conservative response by organized labor. She then goes on to explore the role played by organized labor in state and national party politics.

Racism and middle-class identification, Mink argues, determined the conservative goals and organization of labor: "Ethnic and functional anxieties deflected organized workers away from anti-employer militancy and reform-oriented political action, and toward securing labor-market conditions favorable to trade-union autonomy." Skilled native workers identified more with their employers, with whom they shared religious and ethnic similarities, than with the so-called new immigrants from southern, eastern, and central Europe and Asia. Therefore when they organized, they turned to the craft structure of trade unionism, which befit their ideological nativism, rather than to a more comprehensive industrialism unionism, which would have thrown them together with the new immigrants.

In the Democratic Party native labor found sympathetic allies. Under the leadership of Samuel Gompers and the American Federation of Labor, native labor became an important part of the Democratic Party, but never the most important part. As Mink astutely observes, the "AFL-Democratic ties arose from union-party affinities, chiefly with respect to race, immigration, and the role of the state. The union-party ties expressed union dependence on a middle-class party, not party dependence on a significant labor wing."

As for the new immigrants, before the 1930s they had few political options. They were largely "disorganized at work and disempowered in politics." Even Eugene Debs's 1912 Socialist Party conformed to trade union nativism and ignored new immigrants and native blacks.

In most areas, Mink is remarkably successful. She adds much to the dialogue started by Samuel Lubell, Allan J. Lichtman, Thomas J. Pavlak, and John Allswang. And her discussion of West Coast political nativism is particularly insightful. Her discussion does, however, slight the diversity and political activities of the new immigrants, a group too diverse to be considered as a monolithic entity. Scholars will welcome *Old Labor and New Immigrants*.

RANDY ROBERTS

University of Houston
Texas

MORMINO, GARY ROSS. *Immigrants on the Hill: Italian-Americans in St. Louis, 1882-1982*. Pp. xi, 289. Champaign: University of Illinois Press, 1986. $21.95.

This is a better than average local history describing one group of immigrants who settled in one American city, worked and prospered, and finally were more or less assimilated. It deals with Saint Louis rather than an East Coast city, and with North Italians rather than the more typical migrants from Sicily, Naples, or Calabria. After an introduction, successive chapters describe nineteenth-century Saint Louis and the Lombard town from which most of the migrants came; the formation of the new community in Missouri; how Prohibition affected it; church, politics, and recreation; life during World War II; and, in a brief epilogue, how the community fared up to 1982. A 21-page bibliography is followed by an adequate index.

The wide range of sources includes oral histories from both sides of the Atlantic, newspaper reports, official documentation of various types, and commentary from other studies of immigrants. In an apparent attempt to compose a book that would interest the general reader, all of these are jumbled together into a sometimes disconcerting mixture. Local color is provided by an annoying mannerism of furnishing Italian words, each of which is immediately translated into English between parentheses. Similarly, bits from the oral histories are reproduced in what Mormino imagines is an authentic rendering of broken English. Such passages are especially unfortunate because, when he is not striving for an effect, Mormino can produce a clear and sometimes even elegant exposition.

Until a streetcar line was built in the 1920s, the Hill, as the Italian quarter was called, was isolated from downtown Saint Louis, and this physical insulation was reinforced by a high incidence of home ownership, intergenerational continuity in occupation, and a protracted uninvolvement in municipal politics. How self-contained the community still was in 1915 is indicated by an interesting comparison of Italian businesses with those run by Greeks. Of a total of 402 Greek establishments in Saint Louis, 140 were restaurants and 94 were confectioneries—both obviously catering to a general clientele. In contrast, Italians ran Italian groceries and saloons that served the same local function as English pubs; Italian-run fruit stores, which were distributed throughout the city and thus contradicted the usual pattern, were owned by Sicilians rather than North Italians. When the isolation started to break down in the 1950s, the Hill was a kind of museum piece, excellent for a historical study but for that very reason perhaps atypical of the life elsewhere of third-generation immigrants.

Much of the book is detailed description. Two of the more analytical passages both try to explain a negative—why upward mobility was slower among these immigrants and their descendants than among some other nationalities, and why the Italians took so long to use the political machine of Saint Louis to further community interests. The analyses are good enough to make one wish that Mormino had asked why—or why not—more often. To me, the most interesting chapter was "A Still on the Hill," covering how Prohibition was unenforced among these wine-drinking Italians. When saloons were closed, soft-drink parlors opened up in their place. Stills were set up—so it seems from the description—in almost every home, and for their customary cut police officers arrested only those who did not pay. The neighborhood profited from the new enterprise, but few used the opportunity to become big shots and move away. Like everything else on the Hill, crime also was organized to the benefit of the whole community.

For anyone interested in urban history or in ethnic communities, this volume could be a stimulating source. Based on a conscientious gathering of data, the history is told in this work as a story, but I do not think Mormino's main talent is literary. The book would have been improved if the scholarly apparatus had been used to the full to weigh the probably accuracy of old people's memories, yellowed newspaper accounts, and even government reports.

WILLIAM PETERSEN

Carmel
California

OKUN, MITCHELL. *Fair Play in the Marketplace: The First Battle for Pure Food and Drugs*. Pp. xv. 345. De Kalb: Northern Illinois University Press, 1986. $27.50.

Many people think of the Pure Food and Drug Act of 1906 as the beginning of the American consumer movement. But the first federal regulation of a domestic food product was the oleomargarine law of 1886, and before that, in 1881-82, New York, New Jersey, and Massachusetts had passed bills that regulated the adulteration of food and drugs. *Fair Play* chronicles these little-known efforts between 1865-1886 to protect food and drug consumers. The bulk of the book introduces the reader to a complex story of forces at work that faced the difficult problems, beyond filth, of what was adulteration in the new industry of manufactured food and drugs. Did the addition of nonharmful matter, or "sophistication" as it was known—such as color to oleomargarine—constitute adulteration? As Okun tells the story, the answers lay in the interaction of three separate groups that attempted to influence both federal and state legislation and public health boards. Emergent scientists developed various views, which were often tainted by flawed science as well as by industry sponsorship. Passionate gadfly citizens, in a premodern Ralph Nader style, also emerged to carry the fight in the states and in Washington. But the ultimate victors in the three state battles and in the oleomargarine law of 1886 were the conservative food manufacturers and purveyors themselves, who won limited definition and weak enforcement of the adulteration laws. What they sought was "fair play in the markets" in the fashion of Adam Smith, which would by itself ensure consumer protection. It did not happen that way, of course.

This book does not read easily. The narrative is often broken and difficult to follow. Nevertheless, the effort is rewarding and the theme of fair-play conservatism clearly elucidated. The book rests on impressive research in a variety of primary sources. Okun, who basically sees his work as highlighting the persistent conflict in reform processes, shrinks from going beyond his own chronological story and tying his findings to the larger questions of American reform, particularly to the forces that produced American progressivism (1901-17) and the modern post-New Deal regulation of products. Clearly, though, his work supports those historians who argue that later twentieth-century reform is at heart a conservative thrust.

JAMES A. HODGES

College of Wooster
Ohio

SCHRECKER, ELLEN W. *No Ivory Tower: McCarthyism and the Universities*. Pp. viii, 437. New York: Oxford University Press, 1986. $20.95.

During 15 years following World War II, America expelled past, present, and suspected Communist Party members, as well as an array of fellow travelers and otherwise undesirable characters from its universities. Ellen Schrecker depicts this academic purge in a wide range of institutions, focusing on the activities of academics themselves. Her meticulous work, based on a huge set of interviews and correspondence with victims of the purge and on university and government archival data, concentrates on individual cases, to portray idiosyncratic institutions and circumstances. But in the intricate details of individual university proceedings she reveals nationwide mechanisms of subtle coercion that channeled the power of federal and state investigating committees, through university boards of trustees and governors, into the conduct of academic affairs.

Disturbing conclusions emerge that should be pondered at length. Universities, of course, had mechanisms for keeping out undesirables with sound academic credentials long before McCarthy, but national guidelines for tenure protection had made secretive hiring and tenure decisions the major mechanisms for exclusion and expulsion by 1950, so that the true magnitude of the purge and of the blacklist will never be known. Schrecker can therefore detail only cases involving tenured faculty, for whom academic merit was never at issue. Indeed, many victims were leaders in their field. Their being convicted of behavior incompatible with tenure never involved charges that they peddled Communist propaganda in their classes or scholarship. The great majority were long inactive as Communists and most had maintained vague relations with the party even as members, ten or twenty years before. Being a Communist—ever—or being sympathetic with a party position was enough to make a scholar a target. And government, especially in key states like New York, was determined to clean up the universities. Being targeted by or called before a government committee and either testifying or not was the critical turning point for many victims, because the actuality or potentiality of government action triggered university proceedings. Ironically, ideas about academic freedom not only permitted but energized efforts to eliminate the tenured Left, because administrators and faculty alike took that freedom to mean university freedom from state interference more than faculty freedom of speech and association. To protect themselves from public embarrassment or state intervention, universities conducted purges themselves, in their own individual ways, and thus defended their institutional freedom. In this defense, liberals appear to have taken the lead in many cases.

The book has an excellent index; readers can easily find individuals and institutions involved in the purge. The notes and bibliographic essay lead the reader out into the wider world of American anti-Communist hysteria and its historiography. The only real failure is Schrecker's unwillingness to explore the effects of the purge on the substance of American scholarship, teaching, and university life.

DAVID LUDDEN

University of Pennsylvania
Philadelphia

SPERBER, A. M. *Murrow: His Life and Times*. Pp. xx, 795. New York: Freundlich Books, 1986. $22.95.

LICHTER, S. ROBERT, STANLEY ROTHMAN, and LINDA S. LICHTER. *The Media Elite: America's New Power Brokers*. Pp. xv, 342. Bethesda, MD: Adler and Adler, 1986. $19.95.

These two volumes are both concerned with journalists and the world of power politics, but the resemblance pretty much ends there.

The Media Elite is part of a projected multivolume study to examine how social change in America is shaped by competing elite groups. Using sociological techniques and language, it explores its central theme: the elite American media today—the networks, including the Public Broadcasting Service; the *New York Times*, the *Wall Street Journal, U.S. News and World Report, Time*, and *Newsweek*—comprise reporters who largely come from the same context of high-quality universities and who have a progressive or liberal political outlook, which is often projected in their work. Though some of the book's conclusions seem valid, the techniques used in this volume and the lack of a sense of the history of journalism in America invalidate much of the study.

Lichter, Rothman, and Lichter develop their thesis using social science methods such as reactions to vague photos and the coding of the elements in news stories. They proclaim that the personal feelings of the scientists have been eliminated because specific rules have been developed and followed, but they never concede that the very best journalists are able to do exactly the same thing when they report the news.

And though it is quite true that these journalists have often received excellent educations at the best American universities, the fact is that the leaders of American journalism today, the network anchors and the editor of the *New York Times*, did not receive that education. They are instead farm boys from Texas and North Dakota, a high school dropout from Canada, and a refugee from Nazi Germany. And they set the tone.

The lack of a sense of history hurts this volume when Lichter, Rothman, and Lichter infer that the involvement of journalists in public policy is something new in America. It is not; rather, it is a tradition as old as the nation itself. If anything, there is less of it today than at any time in the past.

The reasons for this come forth in Sperber's truly outstanding biography of Edward R. Murrow. A national hero because of his reporting from London before and during World War II, Murrow went on at CBS Television to tackle the toughest subjects in American society. During the 1940s and 1950s he became and today remains a symbol for excellence and integrity in television journalism.

Nevertheless, he found himself being intellectually strangled at CBS, his independence curtailed, his importance more a burden than an ornament. Murrow had put journalistic independence above the needs of the corporation, and he was not to be forgiven for this transgression.

Today's journalists, despite their political inclinations, simply do not have the freedom implied in *The Media Elite*. They are constrained both by the rules of their profession and by the needs of the corporations that employ them.

FRED ROTONDARO

National Italian American
Foundation
Washington, D.C.

STUBBING, RICHARD A. *The Defense Game: An Insider Explores the Astonishing Realities of America's Defense Establishment*. Pp. xv, 445. New York: Harper & Row, 1986. $21.50.

The military-political-industrial complex is so large that no one author can grasp all of its elements and problems. As a result, each of the many recent books criticizing the U.S. defense establishment tends to play the same role as one of the blind men who attempted to describe the elephant. Each book presents an incomplete picture, and it takes several books to get a comprehensive view of the many problems we face in our efforts to defend the United States.

The Defense Game by Richard Stubbing is a valuable addition to this growing body of defense critiques because it is written from the point of view of the Office of Management and Budget. *The Defense Game*, however, is actually two books. The first is a 250-page discourse on the problems of budgeting for, and then purchasing, an adequate defense capability. The second is a 150-page book that covers the careers and accomplishments of all the secretaries of defense from McNamara to Weinberger.

Stubbing's emphasis on the secretaries of defense is appropriate given his contention that the secretary of defense is a pivotal figure in the problems he describes. The examples provided in the book, however, tend to emphasize the politics of defense as the root cause of most problems. All other factors are clearly a distant second in importance, and the secretary of defense is only infrequently mentioned in the first part of the book.

The Defense Game is fascinating to read, but it uses so many lengthy stories to illustrate its points that it tends to be somewhat shallow in its analysis of the problems it presents. In fact, the solutions suggested by Stubbing to the problems he identifies are usually unacceptable for the very political reasons that caused the problems in the first place.

The chapters on purchasing goods and services, the military payroll, and defense manpower problems are the best that have been written in these areas and are well worth the price of the book. The rest of the book is entertaining, informative, and accurate. The book is long on problems and short on solutions—there is only a brief chapter on the latter—but *The Defense Game* is important reading for anyone trying to understand our peculiar method of buying defense.

WILLIAM J. WEIDA

Colorado College
Colorado Springs

WALD, KENNETH D. *Religion and Politics in the United States*. Pp. xiv, 301. New York: St. Martin's Press, 1986. $29.95.

Religion and Politics in the United States is nothing less than a full examination of the political impact of religious beliefs, institutions, and practices in pluralist America. In this book, political scientist Kenneth Wald argued that American political culture has its root in religious thought. The leaders' right to rule, limited government, and an acceptance of diversity—three fundamental American beliefs about governance—were originated from the central doctrines of Puritan thought, namely, covenant theology, an image of the sinful human nature, and the concept of a chosen people. The pervasiveness of these shared political values, however, has not eliminated religious sources of political conflict.

Instead, the durability of religion in American society has given rise to persistent patterns of institutional coexistence and political conflicts, a theme that can be inferred from Wald's findings. Within the religious community, Wald found interdenominational disagreements and contention. For one thing, none of the major denominational groups that Wald examined—Catholic, Jewish, black Protestant, white mainline Protestant, and white evangelical Protestant—enjoys a clear majority among the American public. Instead, these groups are divided, to varying extents, over social, partisan, and ideological issues. Policy divergences are deepened by the groups' political involvement through direct public action, lobbying, campaign activities, and seating their own church members in governmental positions. Intense intergroup competition, however, often reduces the substantive impacts of religious influence on the policymaking process. Accommodation and conflict have existed in church-state relations since the times of the founding fathers. During the past forty years, judicial decisions have both widened the proper line that separates church and state and, at the same time, have remained committed to individuals' rights to practice their faith.

The success of institutional accommodation in the face of religious conflict ultimately depends on a delicate balance between religious activism and democratic governance. Wald takes up this critical issue in the concluding chapter. On the one hand, democratic institutions may be weakened by aspects of religious practices, such as closed-mindedness, dogmatism, and lack of commitment to accommodation. On the other hand, religion also provides the standards for such crucial democratic concepts as fairness, dignity, justice, and equality that may guard against totalitarian tendencies. Given the inconclusive evidence in both pros and cons,

Wald adopted the middle ground: "that religion in politics is neither an unvarying source of good nor a consistent evil influence."

An informative discussion notwithstanding, the book has paid modest attention to certain issues that may provide a foundation for a more theoretical perspective on the politics of religion. Can one treat religious institutions as just another set of interest groups in the pluralist political system? Or are they primarily organized in terms of ideas—that is, moral values—and only secondarily concerned with tangible interests? Further, has the religious factor, which largely follows ethnic lines, reinforced the territorial organization of political institutions, such as legislatures? In this regard, to what extent have class differences been tempered as a result of the durability of religion in the political process? Above all, does religious influence constitute a distinct category of power apart from economic resources, social status, and political power in contemporary American society? Perhaps some of these concerns should have been addressed in the concluding chapter.

Despite these suggestions, I find this book well researched. Further, Wald has approached this often controversial subject with fair-mindedness and sensitivity. This study offers a timely overview and an informative background to the current debate over religion and politics.

KENNETH K. WONG

University of Oregon
Eugene

WESSER, ROBERT F. *A Response to Progressivism: The Democratic Party and New York Politics, 1902-1918*. Pp. xiii, 328. New York: New York University Press, 1986. $35.00.

Robert F. Wesser's *Response to Progressivism* is a study of that fractious and fascinating organism, the Democratic Party. It focuses upon the role this often divided party played in the political wars in New York State in the tumultuous years between 1902 and 1918. Wesser has emerged as the chronicler of Progressive Era New York. His earlier study, *Charles Evans Hughes: Politics and Reform in New York, 1905-1910*, examined the period from the Republican perspective in years in which a New Yorker, Theodore Roosevelt, held center stage as Progressive president. This current volume focuses upon the arch rivals of those Republican Progressives, complex and contentious Democrats. A third volume on the era is projected, and given the quality of *A Response to Progressivism*, I look forward to it.

The story Wesser has to tell is a fascinating one. New York Democratic politics in the first two decades of this century was an exciting world. The upper-level leadership was made up of two richly attractive men who stood in sharp contrast to each other. The middle- and lower-level leaders were also quite interesting, not only because of who they were but because of who and what they would soon be. The constituencies the party leaders dealt with were also fascinating, as portrayed by Wesser, for here too the contrasts in social class, in home environment—rural or urban, upstate or down—were startlingly clear.

At the very center of this world stood two very different men, men whose careers and ideologies influenced the very flow of events. New York City was represented by Charlie Murphy, nominally the leader of Tammany Hall but a man whose influence extended well beyond the borders of his native Manhattan. He was, as portrayed by Wesser, a complex man of enormous charm, of great persuasive abilities, and of a strict morality who was little tempted by the possibilities for gain that his position suggested. He was Irish and Catholic and thus sensitive to the needs of his various ethnic constituencies. Opposite him stood Thomas Mott Osbourne, the patrician, reformist leader of the upstate Democrats. The relationship—beautifully portrayed—between these quite alien beings provides a central focus for Wesser's finely wrought narrative.

Other figures, too, stand out and Wesser, who shows himself a master at the art of portraiture, depicts these men as they enter the teeming world of Democratic politics. Thus we see unfold the embryonic periods in the careers of men such as Alfred E. Smith—by the book's end he is governor of the state—Franklin D. Roosevelt, and Robert Wagner, Sr.

Wesser's central concern is the process of political change, and he focuses upon the way the Democrats redefine their agenda—to put it more in line with the Progressive thrust of the Wilsonian presidency. He also examines the way these leaders at the same time adjust their party's programs to the needs of the Democratic constituency, a constituency itself in the process of change, change wrought by the immigration that created New York.

Wesser demonstrates that he is comfortable in the world of the political scientist and handles demographic and voting-behavior data with ease and with grace, but his fortes are quite clearly biography and rather classic narrative history. He is able to draw any number of satisfying portraits in miniature of the giants he is dealing with and is able to give the reader a sense of the drama and flow that were at the very heart of New York politics in those years. His conclusions are judicious and clearly stated, and he demonstrates throughout a deep knowledge of the sources he is working with. This book is, then, an admirable work in the field of political history, one that can be read with profit by all those interested in America in the twentieth century.

MURRAY A. RUBINSTEIN
City University of New York

WYNNE, LEWIS NICHOLAS. *The Continuity of Cotton: Planter Politics in Georgia, 1865-1892.* Macon, GA: Mercer University Press, 1986. Pp. viii, 200. $22.50.

Lewis N. Wynne effectively traces the planters' aggressive pursuit of economic, social, and political domination of Georgia from 1860 to 1892. Although the Civil War and emancipation seriously struck at the foundation of their power, planters quickly moved at the constitutional convention of 1865 to reassert their strength. Obviously regarding emancipation as a calamity, they wrote a constitution that retained as many of the forms of slavery as possible. Former slaves were left with freedom and little else. A major planter concern was a subordinate laboring class. Contrary to popular opinion, Wynne claims, planters actually preferred a money wage system, but sufficient money was simply unavailable. Planters opposed renting land to blacks. It gave them too much independence. They found an exploitative sharecropping method, which allowed supervision and preserved the prewar class structure, more acceptable.

Planter hegemony was temporarily threatened by congressional Reconstruction when Republicans tried to form a coalition of blacks, mountain whites, poor whites, and industrialists. Unfortunately for blacks, the Republican experiment was easily defeated by Ku Klux Klan violence, intraparty strife, white domination of black employees, and Democratic racism. Probably a greater threat than Republicans to planter control was the New South movement led by Joseph E. Brown and Benjamin H. Hill. Hill and Brown, largely through the Atlanta *Constitution*, urged Georgians to forgo the past and concentrate upon developing factories, railroads, and businesses. Farmers, they said, should diversify rather than depend upon cotton. Racism was less effective against the New South advocates who were natives, often former slaveholders, and had little concern for blacks.

Planters were not unalterably opposed to all industrialization. It was a question of what type and who controlled it. They feared the growth of large corporations and cities that could dilute their power. They favored the formation of many local industries tied to agriculture, and they often invested in such enterprises. Their plan was to "take the cotton mill to the cotton fields." Better a thousand mills scattered throughout the state where they would be subject to planter influence than a hundred mills in Atlanta and Macon. According to Wynne, the major political power struggle in the early 1870s was between planters and the industrial-merchant advocates. For a time the New South movement seemed to be gaining power. The Republican constitution of 1868, which permitted state support of industry and transportation, was to their advantage. Planters by 1875 began a move for a new constitutional convention to regain their power.

Such a convention was held in 1877. The new constitution, in Wynne's words, "reestablished the planting class as the most potent political force in the state, and the restrictions placed on business subordinated that segment of the economy to planter control." The industrialists then joined the planters, whom they could not defeat, as junior partners in an entente cordiale, which, with minor changes, ruled Georgia for the next century.

The Continuity of Cotton is readable and provocative. If Wynne occasionally overstates his case, he does it persuasively. At times the book's organization is slightly confusing, but that is a minor complaint. Overall it is well done and should be of interest to anyone concerned with post-Civil War politics and political power in Georgia.

JOE M. RICHARDSON
Florida State University
Tallahassee

SOCIOLOGY

BETZIG, LAURA L. *Despotism and Differential Reproduction: A Darwinian View of History*. Pp. xi, 171. Hawthorne, NY: Aldine, 1986. $24.95.

JANOS, ANDREW C. *Politics and Paradigms: Changing Theories of Change in Social Science*. Pp. ix, 183. Stanford, CA: Stanford University Press, 1986. $28.50.

Using Thomas S. Kuhn's idea of paradigm shift, Andrew Janos traces the evolution of modern social science in general and political science in particular. He argues that the "classical paradigm" of the social sciences derived principally from the work of Adam Smith and Karl Marx. According to this paradigm, social change is propelled by mankind's progressive mastery of the physical environment by means of technological innovation and scientific discovery. This progress produces changes in economic organization and social attitudes that force adjustments in the patterns of political authority, institutions, and behavior. Despite their differences, both Smith and Marx focused upon social change in polities within the Western tradition, and both viewed political arrangements as generally dependent upon the organization of the economy.

Against this background Janos takes us on a journey through the theories of political and social change of such major figures as Spencer, Comte, Tönnies, Durkheim, and Weber, all of whom he characterizes as having refined this classical paradigm. Nonetheless, anomalies arose when social theorists attempted to apply the paradigm to non-Western societies. Veblen, for instance, pointed out that by borrowing the latest technologies "backward" nations could avoid certain social and economic impediments to progress that innovators in Occidental nations originally had to face. Lenin, too, argued that nations like preindustrial Russia might leap ahead by telescoping their bourgeois—agricultural and commercial—revolutions with the proletarian revolution that would follow the rise of industry. And influenced by Talcott Parsons's multidimensional characterization of the social system, numerous "systems" theorists abandoned unilinear views of development and began to produce theories that suggested alternative paths of modernization that led to democracy, fascism, communism, or political stagnation and decay.

All of these theories failed to provide satisfactory accounts of social change in non-Western nations in the post-World War II period, however. Among other things, they tended to focus on developments within a single country's social and political culture, and they tended to view industrialization as the highest stage of modernization. By the 1960s the situation was ripe for a theoretical breakthrough, a scientific revolution that would produce a new research paradigm.

The new paradigm, while not fully developed, nonetheless has changed the theoretical focus from social systems within nations to a global system that spans nations. Where the old paradigm treated class interest mainly as a function of position in a nation's social hierarchy, the new paradigm considers a larger, global hierarchy. Where the old paradigm looked at differentiation within societies, the new paradigm focuses upon differentiation of the world between simple and complex societies. And the new paradigm also adds a cultural dimension: where the old paradigm emphasized the casual nexus between technological innovation and social consciousness, the new paradigm suggests that ideas themselves can spread across political boundaries regardless of a nation's level of development. A key concept is the international demonstration effect, a refinement of Veblen's emulative imitation. The international demonstration effect leads to attempts to imitate the material standards and life-styles of the West faster than the technological and institutional innovations required to sustain them can be spread.

Janos traces the origins of the new paradigm to sources as diverse as neo-Marxists like André Gunder-Frank, Harry Magdoff, and Immanuel Wallerstein to neo-Weberians like Barrington Moore and Samuel Huntington. He sees its advantages over the old paradigm to include more satisfactory explanations of the differentiation of the class structures of central as opposed to peripheral states; the rise of extractive militaristic states at the periphery; the equalization and effective class alliance between the local elites of peripheral states and those of advanced industrial societies; and the rise of a postindustrial culture in advanced industrial nations.

In his discussion of scientific revolutions, Kuhn makes the point that the revolutionaries are likely to be newcomers to the established discipline, investigators who have invested little in pursuing questions of "normal science" suggested by the dominant paradigm. Where Janos searches for a new paradigm arising from innovators in the fields of political economy, political science, and political sociology, Laura Betzig, a political anthropologist and something of a newcomer, brings to bear an entirely different paradigm, based on the ideas of Charles Darwin, in order to explain the rise of despotism in both Western and non-Western polities.

Defining despotism as "the exercised right of heads of societies to *murder their subjects arbitrarily and with impunity*," Betzig proposes to explain the phenomenon by testing the Darwinian hypothesis that power is exploited to the end of reproduction. Simply put, this hypothesis suggests that despots—men in virtually all cases—have consistently exploited their power to produce more children for themselves and their kin than can their less fortunate subjects. To test this hypothesis, Betzig reviews data on 104—of 106—autonomous or semiautonomous societies available from the Cumulative Cross Cultural Coding Center of the University of Pittsburgh. The data span societies from the ancient Babylonians (1750 B.C.) and Hebrews (621 B.C.) to the twentieth-century Irish (1932), Turks (1950), and Siamese (1955). Using these and other sources she classifies the societies by degree of despotism and by level of polygyny.

Betzig's analysis shows that as the power of men in authority grows, so does their right to their subjects' women, the number and youth of those in their harems, and their ability to seclude them. The Darwinian hypothesis explains preindustrial social structures far better than Marx's notions of primitive egalitarianism or Durkheim's arguments that law served to penalize offenses against the collective consciousness. Nor do theories stressing economic advantage account for the taking and cloistering of prepubescent girls to wife. One is hard put to offer a better explanation than the guarantee of paternity that the seclusion of virgins provides. And while the exploitation of power in the resolution of conflicts of interest in developed societies may also be explained by Marxism or by other theories stressing economic advantage, the Darwinian explanation of powerful individuals seeking to maximize reproduction still holds.

Although Betzig's subsequent statistical demonstration is flawed by an inappropriate use of least-squares correlations on ordinal data, cross-tabulations of the raw data she includes clearly support her interpretation. Indeed, if one were to raise a quibble, it would be only that her speculations about the political implications of her findings are too timid. For example, if the tendency toward despotism were a heritable trait, would it be any wonder that liberal democrats generally find themselves in the minority? And would there be more cause for political theorists to worry about the chances for survival of democratic regimes?

Both of these books are well written, tightly argued, and mercifully brief. They make worthwhile reading for anyone who takes interest in democratic theory.

MICHAEL MARGOLIS

University of Pittsburgh
Pennsylvania

FISHER, SUE. *In the Patient's Best Interest: Women and the Politics of Medical Decisions*. Pp. ix, 214. New Brunswick, NJ: Rutgers University Press, 1986. $25.00.

Fisher writes as both a feminist and a sociologist to "explore the relationship between shared cultural knowledge (norms) and action (medical decision making) to discover how the institutional authority of the medical role is used (its patterned occurrence)." She uses the methods of sociolinguistics to study how physicians and patients at two teaching hospitals communicated to reach decisions about whether to perform Pap smears and hysterectomies. Then she describes how institutional setting, structural context and the "cultural arena" influenced the decisions made in these clinical encounters.

The result is a modest but convincing demonstration that medical power is often used against the best interests of women patients as a result of the way widely shared beliefs are translated into expectations and actions. This is not news, but good scholarship that is also effective advocacy is always welcome.

Fisher's comments on methodology deserve attention from historians, political scientists, and anthropologists. All of us who use qualitative methods will be stimulated by her defense of blending "advocacy with objectivity" and her explication of sociolinguistics. Those who use qualitative methods to study health affairs will be heartened by this demonstration that economists do not have a monopoly on rigorous analysis that has implications for policy.

Fisher fails, however, in her effort to ground her study in the history of medicine. She makes some dubious generalizations and commits errors of fact as a result of inadequate attention to secondary sources, and she sometimes treats the medical profession as monolithic.

DANIEL M. FOX

State University of New York
Stony Brook

FROHOCK, FRED. *Special Care: Medical Decisions at the Beginning of Life*. Pp. xiii, 263. Chicago: University of Chicago Press, 1986. $19.95.

The central issue in neonatology, according to Fred Frohock in this excellent book, is whether medicine should be organized to maintain life at the expense of health. "Do not underestimate the physician's impulse to treat no matter what," Frohock observes, for the "competition to cheat death" is near the center of the normative structure of medicine. And while we may laud in the abstract the impulse to treat, this impulse must be constrained.

The constraints arise from considerations of quality of life and of the harms done to infants in attempting to preserve a life without quality or a life filled with unremitting pain. But such constraints require definitions that are far from universally accepted. To act on a definition of quality is to accept responsibility for the discretionary character of therapy.

The physicians are in a tough spot. They know that other "communities" than the medical one, to use Frohock's term, have interests in therapeutic decision making. There are the parents, of course, some of whom demand treatment despite hopelessness, and others who would treat only the potentially perfect baby. There is the legal community, whose influence is felt by all in the special-care nursery, and there is the religious community, whose currently most vocal elements insist without qualification on the right to life.

Therapeutic decisions must be the outcome of negotiations between different communities. Yet different communities tend to use different languages of evaluation as well as different definitions of humanness. There are, for example, languages of rights, of harms, of interests, of utilities. Frohock carefully compares the rights and harms approaches and finds that a harms language is more sensitively attuned to the dangers created by the decision to treat while also taking account of the potential benefits.

Too often the right to life becomes an obligation to live, and as such bludgeons those most intimately involved—parents, physicians, and infants—into grotesque, hopeless situations. A harms language, which should be congenial to physicians and families alike, considers the physical and emotional pain of treatment as well as the possible deprivation of future good if treatment is withheld.

This book's great value is that it provides a naturalistic account of ethical dilemmas—how they arise, how they affect those involved, and how they are or are not resolved—and does so by capturing passionately and precisely the flow and process of daily events in the nursery.

JOEL S. MEISTER

University of Arizona
Tucson

GLAZER, PENINA MIGDAL and MIRIAM SLATER. *Unequal Colleagues: The Entrance of Women into the Professions, 1890-1940*. Pp. xii, 295. New Brunswick, NJ: Rutgers University Press, 1987. $28.00. Paperbound, $9.95.

This book focuses on the careers of nine women—Mary Woolley, Nellie Nielson, Bertha Putnam, Dorothy Reed Mendenhall, Anne Walter Fearn, Florence Sabin, Alice Hamilton, Mary C. Jarrett, Bertha C. Reynolds—who, during the Progressive Era, forced their way into four professions: academe, medicine, research science, and psychiatric social work. Their stories illustrate the larger process of women's entry into the professions at a time when the professions themselves were just beginning to organize on a national scale. The story Glazer and Slater tell is at once exhilarating and sad: exhilarating because it retrieves part of women's lost history of achievement; sad because so much talent had so little impact on the professions and organizations that these extraordinary women sought to join.

In the particulars of their data, Glazer and Slater discover four strategies their subjects used to cope with the pervasive sexism that confronted them. These were superachievement, separation from male-dominated institutions and the building of female-centered ones, innovation within careers, and, finally, subordination to male colleagues. These strategies were used in various combinations that shifted over the span of a career. Access to professional training often could be secured only by superachieving, while later work was sustained by a judicious combination of withdrawal—for example, into women's colleges—innovation, as in Alice Hamilton's pioneering efforts in workplace health and safety, and subordination, as when women accepted research-assistant positions in laboratories.

Different though these various strategies were, they had one thing in common: they avoided a head-on confrontation with male professions. To be sure, there was enough individual acrimony. But on an institutional level, separation and innovation produced enclaves in which women could pursue their careers without taking positions away from male colleagues. Women's colleges, the medical subfield of public health, the female profession of psychiatric social work, and the ranks of lab assistants became a female turf built by tenacious and farsighted women, but often not wrested from men. The women themselves were aware that this strategy gave them only a marginal and precarious toehold in the realm of the professions. They were particularly frustrated by their inability to protect protégés and to create scientific progeny.

Their story, then, is one of heroic struggle, great personal vindication, and little institutional change. Is it fair to have asked more of these women? Glazer and Slater touch on this point lightly. They mention but do not elaborate on the fact that their subjects challenged the gender politics of their day largely by calling on its class politics—that is, by wrapping themselves in the banner of progress through scientific, authoritative knowledge applied to social betterment by a highly trained elite. The limited success of their enterprise raises the question of whether one type of egalitarian politics can succeed in isolation from others. Does gender politics also require class politics? Glazer and Slater leave us to draw our own conclusions from the lessons of history. They have provided a vivid and insightful account to inform our considerations.

EVE SPANGLER

Boston College
Chestnut Hill
Massachusetts

HOLLINGSWORTH, J. ROGERS. *A Political Economy of Medicine: Great Britain and the United States*. Pp. xix, 312. Baltimore, MD: Johns Hopkins University Press, 1986. $37.50.

Hollingsworth's well-balanced study of medical care in Britain and the United States will justifiably occupy an important place in the growing body of comparative literature on welfare state services in the modern democracies. The study is divided into three parts: the structure of each health delivery system, an analysis of system performance in a comparative setting, and a short concluding section. Unlike many more narrowly conceived policy comparisons, Hollingsworth gives us a reasonably good, if highly compressed, historical account of how the two systems developed. The interesting problem is how to establish the validity of the performance data, which occupy much of the book, with the analytical conclusions that are largely based on group politics.

The first section depends heavily on existing histories of medical care in the two countries, but it still provides a useful account of how the two systems developed. One wishes that Hollingsworth had had time or inclination to use more extensively some of the British cabinet papers now available from the Public Records Office. Much of the ground he covers can now be examined in much closer detail. In fleshing out the historical context, extensive, if not always reliable, biographies of major figures are also missing. A lot is known about why the two systems developed as they did, which the external evaluation of the two systems fails to tell us.

In failing to do so, it makes comparison in a policy sense more difficult.

This difficulty becomes clearer in the second section of the book, where Hollingsworth claims to be engaged in comparative analysis. To be sure, there are some good tables showing different aspects of performance in relation to such general goals as equality of care and so forth. What one lacks is any contextual knowledge that helps us understand why, for example, the British seemed to think equality of care more important than did the United States, or at least seemed to think that the National Health Service, for all the later second thoughts, was the way to achieve equality. The approach creates the illusion that there was a positive consensus about the National Health Service whereas in fact there was surprisingly little serious political debate outside the charmed circle of physicians and government officials who actually made the key decisions. Bevan himself, like the Edwardian social reformer Lloyd George fifty years earlier, was prepared to make enormous concessions to the private hospitals and to the research concerns of the medical elite in order to smooth the way for his bill in Parliament.

Some of these comparative difficulties become more apparent in the concluding section, where Hollingsworth tries to integrate existing theoretical literature, much of it about group behavior, with his findings about the two health systems. What is less clearly portrayed is that group politics in the two countries may mean fundamentally different things. Whether one prefers Alford's somewhat Marxist view of group behavior or conventional pluralist views of group politics, the problem remains of showing that these models have similar meanings across political systems. Nonetheless, this book provides valuable information for the comparative policy student and will no doubt be widely used in building more satisfactory accounts of the two very different health systems.

DOUGLAS E. ASHFORD
University of Pittsburgh
Pennsylvania

LEES, ANDREW. *Cities Perceived: Urban Society in European and American Thought, 1820-1940.* Pp. xi, 360. New York: Columbia University Press, 1985. $30.00.

This study provides the reader with an overall survey of European and American reaction to the urban phenomenon during the nineteenth and twentieth centuries. The book chronicles the varied opinions of politicians, businesspersons, social workers, and intellectuals regarding the development of urban centers within their societies. It faithfully records their positive and critical evaluations and endeavors to search for the origins of these sentiments and to assess their broader implications for the future development of their respective industrial nations.

Lees begins his inquiry into the development of "urban consciousness" by considering its first manifestations in Europe and North America during the early and middle decades of the nineteenth century. He examines in quick succession the various critiques made of the new urban life-style in Victorian England, France, Germany, and the United States. He cites the views of both well-known and obscure writers and commentators, providing a broadly inclusive picture of the doubts, hopes, and concerns that occupied the minds of those who confronted the conditions of the modern age. In this portion of the book, he stresses the distinct national variations to be found in "urban consciousness" and links these to existing differences in historical, political, social, and economic development.

From this initial overview, Lees directs the reader's attention to the pivotal years of 1880-1918. He asserts that it is in this time period that the most serious reflection on the problems and potentials of urban life were undertaken by the citizenry of Europe and the United States. In the course of five chapters he sets forth their commentary. He apportions time equally between critics and proponents of urban life. This time, however, he considers their indictments and praise from outside a particular country context, arguing that themes such as city as health threat or city as cultural stimulus tended to transcend national borders.

Lees concludes his book with a careful consideration of attitudes toward urban life between the world wars. He notes that the overall commonality in opinions that existed between countries prior to 1918 tended to collapse in the wake of World War I. He suggests that a significant divergence in attitude appeared between Europe and the United States. Germany, in particular, tended to develop unique, national perspectives on the urban condition. Lees links these growing differences in opinion to the dissimilar economic, political, and social forces faced by each society.

There is much that can recommend Lee's efforts. First, his book is one of the few that has endeavored to focus on the reaction to urbanization and not simply on the forces that brought it about. Second, it is well researched and amply illustrated and footnoted. Its bibliography attests to the thoroughness of Lees's inquiry. Third, the comparative perspective that Lees utilizes through most of the book highlights common themes that might have escaped a single-nation study.

There are some limitations to Lees's investigation, however. The first relates to his selection of countries to be studied. He does not provide an adequate explanation for why Britain, France, Germany, and the United States are to be the focus of the inquiry. Why are smaller European states—such as Sweden or the Netherlands—Russia/USSR, and Japan excluded? Conceivably they would provide equally interesting case studies as well as a valuable check on generalizations regarding the commonality of opinions between nations in the 1880-1918 period. Second, and perhaps more disturbing, is the lack of extensive conclusions drawn by Lees from his study. It is only in the final few pages of the book that he attempts to account for the variation in attitudes displayed by the different observers he has quoted. He links these differences ultimately to class, ideology, and "rhythmic alterations of sentiment." In each case, additional evidence and analysis need to be offered.

Such flaws, however, should not detract from the overall value of Lees's work. *Cities Perceived* provides a useful addition to the existing literature in urban studies. It provides a good intellectual history of the urban phenomenon from a comparative perspective. The book will no doubt serve as a base from which additional investigations of historical and contemporary reaction to the urban condition can take place.

DOUGLAS C. NORD

University of Minnesota
Duluth

PIELKE, ROBERT G. *You Say You Want a Revolution: Rock Music in American Culture*. Pp. 270. Chicago: Nelson-Hall, 1986. $23.95.

As is the case with all modern media, rock music's analysts tend to be devotees rather than critics, so the field remains wide open for sensible histories of this important moment of political upheaval and aesthetic change. Robert Pielke's "holistic, philosophical appraisal" attempts to fill the void by quite rightly placing musical change at the center of recent cultural tensions.

Believing that "a truly revolutionary force is at work in our culture," and that rock 'n' roll was and is "the principal means for the dissemination of revolutionary values," Pielke traces 1960s' rebellion back to the duck walk and the scandalous Little Richard. Middle-class youth's attraction to taboo so-called Negro music in the early 1950s coalesced in Elvis Presley, a more palatable symbolic fusion of white disillusionment and black struggle into a subversive "negation" of the status quo. The requisite "affirmation" of insurgent new values can be seen in the creative energies of The Beatles, who helped proliferate the search for greater freedom, individuality, mutuality, and harmony in the following decade.

Pielke insists that the process continues today despite the fabled flight to the right. Evidence of "revolutionary values" can be seen throughout the malaise of the 1970s and even the reactionary politics of the 1980s, for additional marginal groups—women, gays, southerners, and Latinos—have asserted their alternative social views in a musical idiom that Pielke sees as no less powerful than religion.

Unfortunately, too much revolutionary cant and too little convincing analysis circumvent this work from the rich vein of its topic. Pielke has marshaled Sartre, Tillich, Maslow, and, of course, Marcuse to convince us that the times they are a-changing. All of this is unnecessary, as his basic points are quite straightforward. Relying upon Marshall McLuhan's questionable authority, he insists that the very form of rock is inherently "revolutionary," so any helpful interpretation of the music itself bows to abstractions over the "medium." The possibility of ambiguous content is dismissed rather glibly as well.

The best writing involves The Beatles and the 1960s, in which a direct discussion of people and events yields some valuable insights. Here Pielke touches upon the important irony of countercultural icons becoming commodities themselves for a mass society that they seemed to oppose. He also begins to use his philosophical talents to unravel some of the counterculture's politics.

However one may regard Pielke's self-conscious radicalism, he is essentially correct about rock 'n' roll; we cannot overestimate its significance, especially for a generation and time that were looking beyond orthodox boundaries for spiritual and political values. But calling it "revolutionary," while it may be true in the broadest sense, does not do justice to the complexity of the art or history. A fine opportunity has been missed here to use the music itself, its lyrics, cadence, and powerful social role, to illuminate a remarkable American moment. The importance of rock 'n' roll is that it embodied a rare, critical merging of politics, social tension, and art.

Moreover, rock was a culmination of several social and aesthetic trends that had been evolving throughout the modern decades, a historical perspective that is sorely lacking here. Considering his politics, it is surprising that Pielke never consults the radical Americanists, Warren Susman, Jackson Lears, or Christopher Lasch, who offer a promising foundation for discussing how a socially disruptive popular medium can become absorbed by, even cooperative with, the mainstream.

In its general outline, *You Say You Want a Revolution* correctly invites us to approach rock 'n' roll seriously and be aware of its social implications, but the argument is seriously hampered by overstatement and an insistence upon discussing the medium in a most abstract fashion rather than the message in detail. A convincing and thorough synthesis of the rebellion's politics and culture remains unwritten. Until then, the ethos of the period is evoked best by Theodore Roszak's *Making of a Counter Culture*, which Pielke considers dated, while the standard critical history of the music remains Charlie Gillett's *Sound of the City*, which Pielke does not cite in this study.

STEVEN SMITH

Brown University
Providence
Rhode Island

SCOTT, HILDA. *Working Your Way to the Bottom: The Feminization of Poverty*. Pp. xii, 192. Boston: Pandora Press, 1985. Paperbound, $8.95.

SIDEL, RUTH. *Women and Children Last: The Plight of Poor Women in Affluent America*. Pp. xviii, 236. New York: Viking Penguin, 1986. $16.95.

One of the most important scholarly collections to emerge in the past two decades is the gender literature, particularly the feminist. Similar to the efforts by black scholars to correct a white bias in scholarly work, the feminist literature attempts to correct a male bias in scholarly writing. Indeed, this effort is not just a matter of correction; it also involves assuming perspectives and addressing issues not previously covered.

These two volumes represent the best in

feminist literature as they address an issue of fundamental importance, the economic exploitation of women in America and the world over. Hilda Scott's is a tight, well-written book that reads like a high-quality professional report. Scott is a writer and journalist who has written about women and the economy, focusing on the comparative study of women in different types of industrial societies. For her, the "feminization of poverty" refers to "the whole complex of forces that keep women in an economically precarious situation—considerably more precarious than that of most men—while increasing their economic responsibilities." Much of feminine poverty is invisible because women perform 90 percent of the world's unpaid labor. This is because women are still largely restricted to work conforming to a traditional gender division of labor. Their so-called natural role still largely confines women to work of low social and economic value. Scott suggests economic policies creating alternatives to the two main sources of women's poverty, "their unpaid labor in the family and their consignment to jobs in a female ghetto of poorly paid work."

Ruth Sidel's book is a study of the causes of and policy solutions to the feminization of poverty in America. Sidel is professor of sociology at Hunter College of the City University of New York and a specialist in family problems and policy. Among the causes of this development she lists the growth of female-headed families, a dual labor market that actively discriminates against female workers, a welfare system that seeks to maintain its recipients below the poverty line, the unpaid domestic responsibilities of women, and a Washington administration that is reducing funds for the needy.

Utilizing personal interviews, reports, and public documents, Sidel points to a bimodal development in America: that as the number of general poor has decreased, the number of poor women and children has increased. The startling fact is that two-thirds of Americans below the poverty line are women. Although the structural causes are complex, many poor women are not born poor but descend into poverty as a result of such events as separation or divorce, death, or birth of a child.

Drawing upon her knowledge of China and Sweden, Sidel formulates policy recommendations for American society. These include "paid maternity and childcare leave, maternal and child health care, children's allowances, a national system of day care administered locally for all families in need, and an aggressive child-support program."

Both of these excellent books are straightforward and unponderous, factual and policy oriented. They are highly recommended for all social scientists, policymakers, community workers, and citizens concerned with the future welfare of America's prime source of human resources, its families.

SEYMOUR LEVENTMAN

Boston College
Chestnut Hill
Massachusetts

STEIN, JUDITH. *The World of Marcus Garvey: Race and Class in Modern Society.* Pp. xii, 294. Baton Rouge: Louisiana State University Press, 1986. $22.50.

The World of Marcus Garvey is an intriguing one. As discussed by Judith Stein it is one that can only be understood as a meld of people, historical events, and social, economic, and political forces. The story of the rise and fall of a relatively obscure individual, and the movement he generated, is told in rich detail. The highlights of Garvey's life are thoroughly described. So too are the ebbs and flows of Garveyism as it reached out to enlist and repel a substantial portion of blacks throughout the world.

In telling this complex story, portraits of the key individuals who joined and/or abandoned Garvey are finely sketched. A good feel for the human side of a social movement results. But Garvey's drama is not just told through mapping the entrance and exit of a cast of interesting characters. History is not forgotten. Garvey and his movement are well situated in the context of his times. The uplifts and downward drafts precipitated by the vicissitudes of politics, economics, and, in general, societies in change are carefully integrated into this account. While the research for this analysis is comprehensive, the documentation of facts is parsimonious and insightful. Detail does not detract from the smooth flow of the narrative.

What is missing from the book is a solid theoretical foundation. Because of this limitation, the book does stumble on occasion. No attention is paid, for instance, to personality theory. Hence a psychologically oriented reader looking for connections between Garvey's organizational drive and the forces that drove him from within will be disappointed. Implicit attention is given to propositions that help explain the growth and decline of social movements. The value of this book for a student of social movements, however, lies more in the factual detail provided than in the undeveloped theoretical presentation of Garveyism as a social phenomenon.

Greater success is achieved in placing Garveyism within an economic framework. Stein stresses—albeit far too briefly—that the vagaries of a capitalist economy both spurred and eventually doomed the movement. It is quite correctly argued that Garvey's business adventures—far and away the cornerstone of his organization—can only be understood as a part of the general dynamics of world economic forces operating at the time. For those with a fancy for theory building, however, the economic scaffolding is only partially completed. A fuller statement explaining the fortunes—and misfortunes—of Garvey-like movements is still needed.

Nonetheless, this book adds not just to the literature on black history. Its fertile historical base should gain it access to the compendium of good works both on social movements and on the political economy of powerless people.

J. W. LAMARE

University of Canterbury
Christchurch
New Zealand

ECONOMICS

BELDEN, JOSEPH N. *Dirt Rich, Dirt Poor: America's Food and Farm Crisis*. Pp. xii, 188. New York: Routledge & Kegan Paul, 1986. $27.50. Paperbound, $12.95.

DUCHENE, FRANÇOIS, EDWARD SZEZEPANIK, and WILFRED LEGG. *New Limits on European Agriculture: Politics and the Common Agriculture Policy*. Pp. xiv, 286. Totowa, NJ: Rowman & Allanheld, 1985. $45.00.

These books present an interesting contrast. Throughout most of human history, agriculture has been the intense preoccupation of an overwhelming percentage of humans. In the twentieth century, however, in three regions—Europe, Soviet Eurasia, and North America—for different reasons, but to similar effect, agriculture became a marginal activity involving only small percentages of the full population.

This shift has reduced the political importance of the agricultural sector, but it has not reduced its economic, material importance. As William Jennings Bryan used to emphasize, agriculture remains the necessary substructure on which the urban civilization depends. People and nations can survive without videocassette recorders and word processors but not without food.

Joseph Belden's book is an activist's manual. He presents facts, statistics, and slogans.

> Farm debt has increased. Farm loan delinquencies are increasing . . . more than 40 [percent] of the [Farmers Home Administration] loans were delinquent. . . .
>
> During the 1970s, foreign demand grew about 8 [percent]. . . .
>
> Farmers in the 1980s face hard times not seen since the 1930s.

New Limits on European Agriculture, by Duchene, Szezepanik, and Legg, is a scholarly study and a research tool. This work focuses on the political aspects of economic and agricultural policy.

> The CAP [Common Agriculture Policy] . . . a fusion of the policies of the member states in a single European system . . . is also . . . a very unfused accumulation of persisting national ambitions and political pressures.
>
> The CAP is an isolated relic of the ambitions of the founding fathers of the European Community.
>
> One clear implication . . . is a juxtaposition of national policies that continue on their own sweet way—all the more because the CAP, by monopolizing attention, draws a veil over what they are and signify.

The common theme is evident. Agriculture has become politically irrelevant because it involves a minimal percentage of the population. It remains economically important, however, because it remains the essential basis for human life and for all other economic activity.

GEORGE FOX MOTT

Mott of Washington and Associates
Washington, D.C.

BENNETT, DOUGLAS C. and KENNETH E. SHARPE. *Transnational Corporations versus the State: The Political Economy of the Mexican Auto Industry*. Pp. xiii, 300. Princeton, NJ: Princeton University Press, 1985. $42.00. Paperbound, $9.95.

Bennett and Sharpe, both political scientists, explore the transformation of the automobile industry in Mexico since 1960 at three levels: (1) a historical account of the growth of automobile manufacturing in Mexico; (2) bargaining between states and transnational corporations (TNCs) in the overall context of the dependency of a developing country, focusing on the possible distortions of a less developed country's economy, society, and politics that can follow from the activities of TNCs; and (3) a demonstration of the ability of a historical-structural method that is neither deterministic nor voluntaristic to address the fundamental issues of social science inquiry.

Dependency and bargaining incorporate the central methodological perspective of this book: the historical-structural approach. Alleging that the dominant approaches in American social science have tended toward either determination or voluntarism, Bennett and Sharpe postulate that each social actor has interests and power of its own, the possibilities for action are limited by the structures in which the actors are "enmeshed," and these structures are historical products of past human actions. In sum, actors can make their own history, but they are also constrained by social structures. This approach is the organizing principle of the entire book. Chapters 2 through 4 introduce the major actors—the TNCs and the Mexican state—and the structures within which they are enmeshed, such as the number and ownership of firms in the auto industry for the former and the "embedded orientations" acquired and institutionalized in the course of history for the latter. Chapters 5 through 10 analyze dependency and bargaining in the automobile industry in Mexico between 1960 and 1980, concentrating on three bargaining episodes.

In its first bargaining encounter with auto TNCs, in 1962, the Mexican government sought to create a manufacturing industry in Mexico by means of an import-substitution policy. In its second bargaining encounter, in 1969, the government tried to shift the orientation of the now-established industry from import substitution toward export promotion. In its third bargaining encounter, in 1977, the government wanted to strengthen export requirements. The treatment of these encounters is the major contribution of the book.

Bennett and Sharpe conclude that the interests of the transnational automobile companies caused them to pursue strategies that were injurious to Mexico, but that the Mexican government was able to mitigate many of the negative consequences through the exercise of its own power. This conclusion will strike many readers as bewildering because the criteria used to determine injury appear to be mostly the policy objectives of the Mexican government itself, namely, the creation of an automobile industry dominated by Mexican ownership that would stimulate economic growth without burdening the balance of payments. For instance, we are told that the TNCs were reluctant to invest in Mexico but that the Mexican government made it "rational" for them to begin manufacturing vehicles in Mexico by threatening to deny access to the Mexican market. Again, the TNCs used the same capital-intensive technology in Mexico as in their home countries, produced the same models—too expensive for the poorer part of the Mexican population—and followed product-differentiation strategies with many models and annual model changes. According to Bennett and Sharpe, the failure of the government to alter this "injurious" behavior is traceable to its initial mistake in 1962 of admitting too many manufacturers. Economists, among others, would draw a clear distinction between the policy objectives of the Mexican government and the economic consequences of TNC behavior. The question of whether the automobile policies espoused by the Mexican government were good for the Mexican economy is not seriously addressed in this book.

A second conclusion is noncontroversial. State policy tended to succeed when the structure of the auto industry made the interests of the TNCs consistent with the goals of the state. For instance, the shift to global sourcing strategies by TNCs in the 1970s made them more supportive of the export-promotion policy of the Mexican government.

The dependency perspective, even when adjusted by the historical-structural approach, disposes Bennett and Sharpe to end their study with gratuitous—and tendentious—conclusions about the consequences of the automobile TNCs for Mexico's economic development. These conclusions are excess baggage because they are not derived from the subject of their book, which is an exhaustive inquiry into the bargaining episodes between the Mexican government and transnational automobile companies. On that subject, this book is strongly recommended.

FRANKLIN R. ROOT

University of Pennsylvania
Philadelphia

BODE, FREDERICK A. and DONALD E. GINTER. *Farm Tenancy and the Census in Antebellum Georgia*. Pp. xxi, 278. Athens: University of Georgia Press, 1987. $27.50.

Farm Tenancy and the Census in Antebellum Georgia examines the dual economy thesis, which sees the antebellum South as a region of large plantations and yeoman farmers, but not much else. In the process, it also questions related views that tenancy and particularly sharecropping were productive relations that not only expanded rapidly in the postbellum period but were also qualitatively new.

In order to investigate the extent and spatial distribution of tenancy in antebellum Georgia, Bode and Ginter conduct a detailed examination of the manuscript returns for the 1860 federal census. The authors suggest that previous interpretations of the census are seriously flawed because they take the absence of a category for tenants as implying the absence of tenants themselves. Their analysis of the returns indicates that enumerators dealt with this difficulty by listing crops and livestock under the tenants themselves, but assigning acreages and farm values to their owners, not to the tenants. Hence the frequent occurrence of "farmers without farms," who make up the bulk of Bode and Ginter's tenants.

On the basis of this and a number of other related findings, Bode and Ginter construct four levels of estimates for tenancy rates, ranging from level one, an "indisputable minimum" (p. 115), to level four, the authors' preferred estimates, which indicate county tenancy rates of 20 to 40 percent of farm operators. Bode and Ginter prefer the latter estimates not only due to the evidence collected during their investigation of the census, but also because they appear to be consistent with a variety of hypotheses about the relationship between tenancy, soil quality, the economics of slave production, farm size, and the distribution of personal wealth. Georgia tenants in 1860 were generally poor, operating at or beyond the margins of the cotton economy. Yet they were numerous, and many of them may have been sharecroppers—"the most contentious of [Bode and Ginter's] conclusions."

Farm Tenancy and the Census in Antebellum Georgia is a very thorough and detailed investigation of the data upon which much historical investigation has been based, and it is a model of how to proceed when data are conceptually ambiguous. With three full chapters and portions of others devoted to method, it will appeal primarily to the professional historian rather than his or her armchair counterpart. If its conclusions are correct—and scholars familiar with the 1860 census manuscripts will have to judge—it will challenge a variety of existing preconceptions about the nature of the antebellum South and also about the political economy of the postwar transition.

PHILLIP WOOD

St. Francis Xavier University
Antigonish
Nova Scotia
Canada

COHEN, BENJAMIN J. *In Whose Interest? International Banking and American Foreign Policy*. Pp. xi, 347. New Haven, CT: Yale University Press, 1986. $19.95.

Benjamin Cohen examines the impact of international banking on the making of U.S. foreign policy. He argues that high finance and high politics can no longer be separated. Therefore, he concludes, bankers and government officials need to stop arguing that their activities are independent of one another and learn how to cooperate. He boils down his recommendations into a handful of capsule precepts—"Five Commandments for an Improved Bank-Government Relationship." These are (1) talk together, (2) involve responsibles, (3) think strategically, (4) provide information, and (5) accommodate interests.

There is very little that is new or startling in this volume. It is clearly and accurately written, although sometimes Cohen's determination to keep the style light and easy distracts more than it simplifies. The book is aimed more at students, bankers, and policymakers than at scholars. Cohen's strength is that he synthesizes a great deal of interesting information and sensibly highlights a relationship that should be obvious to everyone but often gets lost in bureaucratic jealousies and the shortsightedness of public and private sectors alike.

Part 1 lays the groundwork by chronicling the rapid growth and internationalization of banking activities during the last "incredible quarter century" and also explores when public and private interests converge and clash.

But what is new? Part 2 starts by comparing past and present international banking practices. Cohen believes that the difference "lies in the details," but that these variations are important. To demonstrate just how important, he examines four case studies that form the heart of the book. Case one considers whether the vast horde of petrodollars accumulated by Arab-OPEC countries during the 1970s influenced U.S. policy. Cohen concludes that although "the purported money weapon was in reality far less of a threat than it appeared to many Americans at the time, it nonetheless worked—in the sense of exercising practical leverage over U.S. government policy." Case two examines how the freezing of Iranian funds in U.S. banks after American hostages were seized in Tehran in November 1979 drew the banks into the political picture. As in the first case, Cohen argues "that large concentrations of assets in the United States under the control of other governments do matter from a foreign policy point of view." In case one, U.S. interests were compromised; in case two they were assisted. Case three assess the extent to which the existence of large Polish debts to Western banks complicated East-West tensions arising out of the suppression of the Solidarity movement. Finally, case four investigates the more familiar terrain related to finance and foreign policy connections inherent in the Latin American debt crisis after August 1982.

In the final section Cohen advances his recommendations for government officials and bankers.

JONATHAN D. ARONSON
University of Southern California
Los Angeles

DIETZ, JAMES L. *Economic History of Puerto Rico: Institutional Change and Capitalist Development*. Pp. xxxiii, 337. Princeton, NJ: Princeton University Press, 1986. $65.00. Paperbound, $20.00.

Surveying the economic history of even the smallest of island nations over several centuries would be a formidable task indeed. In the case of Puerto Rico the difficulties are compounded by its ambiguous status under U.S. rule. This makes virtually obligatory an equally fine sense of political history in order to clarify the economic evolution of the island. While it is not without weaknesses, James L. Dietz has produced an excellent synthesis of an abundant literature on Puerto Rico's economic history from the late eighteenth century through the mid-1980s. The work will be required reading for anyone seriously concerned with Puerto Rican themes, and it deserves a wide readership among North Americans in general. Indeed, one of the most startling contributions of this study is its ability to show just how central Puerto Rico was to U.S. economic and military expansion at the turn of the century, how profoundly affected the island was by Rooseveltian reformism and the World War II era, and how abjectly dependent upon foreign-owned export industries and federal government transfer payments the local economy has become most recently. These analyses have relevance far beyond the interests of Puerto Ricanists and ought to be considered by a much larger readership.

The study is organized in five chapters on Spanish colonialism, U.S. sugar-company control to 1930, the crisis of the 1930s, the origins of industrialization and Operation Bootstrap in the 1940s, and growth and misdevelopment since 1950. Essentially, the work builds on three interrelated bodies of scholarship: the new Puerto Rican social history focusing on the agrarian and social history of the period, roughly speaking, 1750-1930; the literature on the political crisis of the 1930s and the rule of the Popular Democratic Party under Luis Muñoz Marín since then; and the largely economic—Dietz is an economist by training—and demographic studies of Puerto Rico in more recent times. In each case Dietz presents a very thoughtful and critical synthesis of a wide variety of studies and perspectives.

Dietz develops a lucid combination of dependency-based and Marxian analyses to judge both the contradictory success of, and the more nebulous and normative appropriateness of, Puerto Rico's development model under Spanish and then U.S. colonialism. To be sure, many of the weaknesses of dependency theory are present here as well: a fairly superficial and telescopic view of 300 years of Spanish rule; a conceptual reductionism on the question of pre- and noncapitalist modes of production versus capitalist ones, as in Figures 1.2 and 2.1; and the normative, judgmental proclivities of the appropriate-technology or basic-needs schools of dependency theory. Nevertheless, Dietz avoids the more serious of these pitfalls owing to his consistent use of more solid Marxian categories of analysis when dealing with internal class structure, his quite profound knowledge of colonialism and its workings in the nineteenth and twentieth centuries, and his ability to recognize and interpret the reality of development, dependent perhaps but development and change still, so often simply denied or reduced to the poverty-of-progress doctrine of the more extreme dependency theorists. Indeed, the subtlety of Dietz's analysis for recent Puerto Rican development led me to an immediate comparison with *Dependent Development: The Alliance of Multinational, State, and Local Capital in Brazil*, Peter Evans's excellent study of dependent development in contemporary Brazil, high praise in and of itself.

The economic history of Puerto Rico is a subject of real interest to North Americans today, as we warily observe the prerevolutionary agonies of its sister colony, the Philippines. Dietz and Princeton University Press are to be commended for producing a finely crafted, richly documented, and very timely study. The excessively high hardcover price may deter many would-be-purchasers, but the 61 statistical tables and 11 plates offer some recompense, while the more reasonably priced paper edition remains an alternative for both public libraries and college courses concerned with our Puerto Rican colony and its peculiar history.

LOWELL GUDMUNDSON
University of Oklahoma
Norman

GREENBERG, EDWARD S. *Workplace Democracy: The Political Effects of Participation.* Pp. 257. Ithaca, NY: Cornell University Press, 1986. $29.95.

Edward Greenberg focuses attention on a little-known chapter in American industrial history with his study of the plywood cooperatives in the Pacific Northwest. Worker-owned companies are not common in the United States, and he compares them with the better-publicized Israeli kibbutzim and cooperatives in Yugoslavia and Spain.

Greenberg reveals a number of skeletons in the cooperative closet. He confesses that an initial enthusiasm of his for worker democracy has been dampened by his investigations. Particularly damning is his conclusion that the cooperatives are less safety conscious than conventional companies. There are a significantly higher number of serious injuries in cooperative plywood mills than in conventional mills. Moreover, the cooperative ethos is compromised by the hiring of nonmember skilled workers and paying some of them more than what shareholding workers receive.

The drudgery of the job is not mitigated by the fact that a majority of the employees are also employers. One member worker remarked, "It's like being a zombie." Another said, "It can drive you nuts." The alienation that workers in repetitive and tedious tasks feel is abundantly present in the cooperatives.

The sense of community that a cooperative would be expected to have is lacking. Contrary to expectation, workers in the cooperatives were less sociable than workers in ordinary plywood concerns. Greenberg's survey revealed, for example, that considerably more cooperative workers ate lunch alone than did workers in the regular mills. The general impression is that cooperative plywood workers are not a very happy crowd.

So while not abandoning hope for cooperative movements, Greenberg has written a critical book exposing serious problems with worker democracy. Less successfully, he invokes Antonio Gramsci in an attempt to suggest a general theoretical framework for his investigations. His own painstaking and numerous interviews do not gain from the brief diversion he makes into social theory.

This book comes as a valuable adjunct to Christopher Gunn's earlier *Workers' Self-Management in the United States* from the same publisher. It has an outstanding bibliography. Parenthetically, it is pleasant to be able to record that Greenberg believes in footnotes that are more than bare citations, and his notes are entertaining and engaging.

P. J. RICH
Department of Training
and Career Development
State of Qatar

OTHER BOOKS

ALEXANDER, HERBERT E. and BRIAN A. HAGGERTY. *Financing the 1984 Election*. Pp. 430. Lexington, MA: Lexington Books, 1987. No price.

ASPREY, ROBERT B. *Frederick the Great: The Magnificent Enigma*. Pp. xvii, 715. New York: Ticknor and Fields, 1986. $24.95.

BANKS, OLIVE. *Becoming a Feminist: The Social Origins of 'First Wave' Feminism*. Pp. 184. Athens: University of Georgia Press, 1986. No price.

BANUZIZI, ALI and MYRON WEINER, eds. *The State, Religion, and Ethnic Politics: Afghanistan, Iran, and Pakistan*. Pp. xi, 390. Syracuse, NY: Syracuse University Press, 1986. $35.00.

BAXTER, CRAIG et al. *Government and Politics in South Asia*. Pp. 415. Boulder, CO: Westview Press, 1987. $39.50. Paperbound, $19.95.

BAYNHAM, SIMON, ed. *Military Power and Politics in Black Africa*. Pp. 333. New York: St. Martin's Press, 1986. $32.50.

BECK, LOIS. *The Qashqa'i of Iran*. Pp. 384. New Haven, CT: Yale University Press, 1986. No price.

BELL, DANIEL and LESTER THUROW. *The Deficits: How Big? How Long? How Dangerous?* Pp. 142. New York: New York University Press, 1985. $15.00.

BENDIX, REINHARD. *From Berlin to Berkeley: German-Jewish Identities*. Pp. 300. New Brunswick, NJ: Transaction Books, 1985. $29.95.

BIBBY, JOHN F. *Politics, Parties, and Elections in America*. Pp. 377. Chicago: Nelson-Hall, 1987. Paperbound, $13.95.

BINGHAM, RICHARD D. et al., eds. *The Homeless in Contemporary Society*. Pp. 276. Newbury Park, CA: Sage, 1987. Paperbound, no price.

BRAMS, STEVEN J. *Superpower Games: Applying Game Theory to Superpower Conflict*. Pp. xvi, 176. New Haven, CT: Yale University Press, 1985. $22.50. Paperbound, $6.95.

BROWN, LESTER R. et al. *State of the World 1987: A Worldwatch Institute Report on Progress toward a Sustainable Society*. Pp. 268. New York: Norton, 1987. $18.95. Paperbound, $9.95.

BUKOWSKI, CHARLES J. and MARK A. CICHOCK, eds. *Prospects for Change in Socialist Systems*. Pp. 149. New York: Praeger, 1987. $34.95.

BURNS, ARTHUR F. *The United States and Germany: A Vital Partnership*. Pp. xii, 51. New York: Council on Foreign Relations, 1986. $10.00.

CARLSON, DON and CRAIG COMSTOCK, eds. *Citizen Summitry: Keeping the Peace When It Matters Too Much to Be Left to Politicians*. Pp. 396. Los Angeles: Jeremy P. Tarcher, 1986. Distributed by St. Martin's Press, New York. Paperbound, $11.95.

CAUSER, GORDON A. *Inside British Society: Continuity, Challenge and Change*. Pp. 250. New York: St. Martin's Press, 1987. $39.95.

CAWSON, ALAN, ed. *Organized Interests and the State: Studies in Meso-Corporatism*. Pp. 260. Newbury Park, CA: Sage, 1985. $35.00. Paperbound, $14.95.

CHILD, JAMES W. *Nuclear War: The Moral Dimension*. Pp. 197. Bowling Green, OH: Social Philosophy and Policy Center; New Brunswick, NJ: Transaction Books, 1986. Paperbound, no price.

COCKS, GEOFFREY and TRAVIS CROSBY, eds. *Psycho/History: Readings in the Method of Psychology, Psychoanalysis, and History*. Pp. xv, 318. New Haven, CT: Yale University Press, 1987. $35.00. Paperbound, $14.95.

COLE, PAUL M. and DOUGLAS M. HART, eds. *Northern Europe: Security Issues for the 1990's*. Pp. xii, 160. Boulder, CO: Westview Press, 1986. Paperbound, $21.50.

CRUSE, HAROLD. *Plural but Equal: Blacks and Minorities in America's Plural Society*. Pp. 420. New York: William Morrow, 1987. $22.95.

CURTIS, MICHAEL, ed. *The Middle East Reader*. Pp. xx, 485. New Brunswick, NJ: Transaction Books, 1986. $34.95. Paperbound, $14.95.

DAALDER, HANS, ed. *Party Systems in Denmark, Austria, Switzerland, the Netherlands and Belgium*. Pp. 372. New York: St. Martin's Press, 1987. $35.00.

DAUDI, PHILIPPE. *Power in the Organization*. Pp. vii, 338. New York: Basil Blackwell, 1986. $49.95.

DENARDO, JAMES. *Power in Numbers: The Political Strategy of Protest and Rebellion*. Pp. xvi, 267. Princeton, NJ: Princeton University Press, 1985. $25.00.

DIMITROV, GEORGE. *Against Fascism and War*. Pp. 125. New York: International, 1986. Paperbound, $4.25.

DiPALMA, GUISEPPE and LAURENCE WHITEHEAD, eds. *The Central American Impasse*. Pp. 252. New York: St. Martin's Press, 1986. $32.50.

DIVINE, ROBERT A., ed. *The Johnson Years*. Vol. 2, *Vietnam, the Environment, and Science*. Pp. 267. Lawrence: University Press of Kansas, 1987. $25.00.

DOBUZINSKIS, LAURENT. *The Self-Organizing Polity: An Epistemological Analysis of Political Life*. Pp. 226. Boulder, CO: Westview Press, 1987. $37.50.

DUTTON, DAVID. *Austen Chamberlain: Gentleman in Politics*. Pp. 373. New Brunswick, NJ: Transaction Books, 1987. $29.95.

EARL, PETER. *Lifestyle Economics: Consumer Behavior in a Turbulent World*. Pp. 314. New York: St. Martin's Press, 1986. $29.95.

EDINGER, LEWIS. *West German Politics*. Pp. xv, 342. New York: Columbia University Press, 1985. $30.00. Paperbound, $10.00.

ELAZAR, DANIEL J. *Exploring Federalism*. Pp. 335. University: University of Alabama Press, 1987. $28.95.

ELSTER, JON, ed. *Rational Choice*. Pp. 266. New York: New York University Press, 1986. Distributed by Columbia University Press, New York. $30.00. Paperbound, $12.50.

EVANS, JUDITH et al. *Feminism and Political Theory*. Pp. 164. Newbury Park, CA: Sage, 1986. Paperbound, $12.95.

FARR, GRANT M. and JOHN G. MERRIAM, eds. *Afghan Resistance: The Politics of Survival*. Boulder, Co: Westview Press, 1987. Paperbound, $25.00.

FEHRENBACHER, DON E. *Lincoln in Text and Content: Collected Essays*. Pp. x, 364. Stanford, CA: Stanford University Press, 1987. $37.50.

FONER, PHILIP S. *History of the Labor Movement in the United States*. Vol. 7, *Labor and World War I 1914-1918*. Pp. x, 410. New York: International Publishers, 1987. $21.00. Paperbound, $9.95.

FREDERIKSE, JULIE. *South Africa: A Different Kind of War*. Pp. 192. Boston: Beacon Press, 1986. Paperbound, $12.95.

FREEDMAN, ROBERT O., ed. *The Middle East: After the Israeli Invasion of Lebanon*. Pp. xviii, 363. Syracuse, NY: Syracuse University Press, 1986. $29.95. Paperbound, $14.95.

FULLINWIDER, ROBERT K. and CLAUDIA MILLS, eds. *The Moral Foundations of Civil Rights*. Pp. vii, 206. Totowa, NJ: Rowman and Littlefield, 1986. $31.95. Paperbound, $14.95.

GENDZIER, IRENE L. *Managing Political Change: Social Scientists and the Third World*. Pp. xiv, 238. Boulder, CO: Westview Press, 1985. $32.50. Paperbound, $11.95.

GILBERT, MICHAEL, ed. *The Oxford Book of Legal Anecdotes*. Pp. xvi, 333. New York: Oxford University Press, 1986. $18.95.

GOLDWIN, ROBERT A. and ART KAUFMAN, eds. *Separation of Powers—Does It Still Work?* Washington, DC: American Enterprise Institute for Public Policy Research, 1986. $16.95. Paperbound, $8.95.

GOLDWIN, ROBERT A. and WILLIAM A. SCHAMBRA, eds. *How Federal Is the Constitution?* Lanham, MD: American Enterprise Institute for Public Policy Research, 1986. $22.50. Paperbound, $10.75.

GOTTDIENER, M. *The Decline of Urban*

Politics: Politics Theory and the Crisis of the Local State. Pp. 299. Newbury Park, CA: Sage, 1987. $29.95. Paperbound, $14.95.

GOTTLIEB, MANUEL. *A Theory of Economic Systems*. Pp. xxi, 249. New York: Academic Press, 1984. $51.50.

GREAVES, RICHARD L. *Deliver Us from Evil: The Underground in Britain, 1660-1663*. Pp. 291. New York: Oxford University Press, 1986. $29.95.

GUMZ, EDWARD. *Professionals and Their Work in the Family Divorce Court*. Pp. xviii, 161. Springfield, IL: Charles C Thomas, 1987. $27.00.

HANLON, JOSEPH. *Apartheid's Second Front: South Africa's War against Its Neighbours*. Pp. vii, 130. New York: Penguin Books, 1986. Paperbound, $4.95.

HARDEN, VICTORIA A. *Inventing the NIH: Federal Biomedical Research Policy, 1887-1937*. Pp. xiii, 274. Baltimore, MD: Johns Hopkins University Press, 1986. $32.50.

HAWKINS, HELEN S. et al., eds. *Toward a Livable World: Leo Szilard and the Crusade for Nuclear Arms Control*. Pp. lxxiv, 499. Cambridge, MA: MIT Press, 1987. $50.00.

HERNANDEZ, JOSE. *Mutual Aid for Survival: The Case of the Mexican American*. Pp. ix, 160. Melbourne, FL: Krieger, 1983. $11.50.

HIGGS, DAVID. *Nobles in Nineteenth-Century France: The Practice of Inegalitarianism*. Pp. xix, 287. Baltimore, MD: Johns Hopkins University Press, 1987. $30.00.

HIRSZOWICZ, MARIA. *Coercion and Control in Communist Society: The Visible Hand in a Command Economy*. Pp. vii, 226. New York: St. Martin's Press, 1986. $27.50.

HOROWITZ, DONALD L. *Ethnic Groups in Conflict*. Pp. xiv, 697. Berkeley: University of California Press, 1987. Paperbound, $12.95.

HOUGH, JERRY F. *The Struggle for the Third World: Soviet Debates and American Options*. Pp. x, 293. Washington, DC: Brookings Institution, 1986. $32.95. Paperbound, $12.95.

HUND, JOHN and HENDRIK W. VAN DER MERWE. *Legal Ideology and Politics in South Africa: A Social Science Approach*. Pp. 132. Cape Town: Creda Press, with University Press of America, Lanham, MD, 1986. Paperbound, no price.

HUYETTE, SUMMER SCOTT. *Political Adaptation in Sa'udi Arabia: A Study of the Council of Ministers*. Pp. xiv, 201. Boulder, CO: Westview Press, 1985. Paperbound, $19.95.

INGRAM, EDWARD. *National and International Politics in the Middle East: Essays in Honor of Elie Kedouri*. Pp. xviii, 284. London: Frank Cass, 1986. Distributed by Biblio Distribution Center, Totowa, NJ. $30.00.

JENKINS, J. CRAIG. *The Politics of Insurgency: The Farm Worker Movement in the 1960s*. Pp. xvi, 261. New York: Columbia University Press, 1985. $30.00.

JOHNSON, ALLEN W. and TIMOTHY EARLE. *The Evolution of Human Societies: From Foraging Group to Agrarian State*. Pp. 360. Stanford, CA: Stanford University Press, 1987. $39.50.

JOHNSON, D. GALE et al. *Agricultural Policy and Trade: Adjusting Domestic Programs in an International Framework*. Pp. 132. New York: New York University Press, 1986. $25.00.

JOHNSON, PAUL, ed. *The Oxford Book of British Political Anecdotes*. Pp. xviii, 270. New York: Oxford University Press, 1986. $17.95.

KATZ, MILTON S. *Ban the Bomb*. Pp. xv, 215. Westport, CT: Greenwood Press, 1986. No price.

KELLY, IAN. *Hong Kong: A Political-Geographic Analysis*. Pp. 191. Honolulu: University of Hawaii Press, 1986. No price.

KLAYMAN, RICHARD. *The First Jew: Prejudice and Politics in an American Community 1900-1932*. Pp. xv, 175. Malden, MA: Old Suffolk Square Press, 1985. $29.95.

KUMAR, SATISH, ed. *Yearbook on India's Foreign Policy, 1983-84*. Pp. 270. Newbury Park, CA: Sage, 1986. $49.95.

LADD, GEORGE W. *Imagination in Research: An Economist's View*. Pp. 146. Ames: Iowa State University Press, 1987. Paperbound, $9.95.

LAIRD, ROY D. *The Politburo: Demographic Trends, Gorbachev, and the Future*. Westview Special Studies on the Soviet Union and Eastern Europe. Pp. xv, 198. Boulder, CO: Westview Press, 1986. Paperbound, $22.50.

LAL, DEEPAK and MARTIN WOLF, eds. *Stagflation, Savings, and the State*. Pp. xii, 402. New York: Oxford University Press, 1986. No price.

LAWSON, STEVEN F. *In Pursuit of Power: Southern Blacks and Electoral Politics, 1965-1982*. Pp. 391. New York: Columbia University Press, 1987. Paperbound, $14.50.

LEHNE, RICHARD. *Casino Policy*. Pp. 268. New Brunswick, NJ: Rutgers University Press, 1986. $30.00.

LEONHARD, WOLFGANG. *The Kremlin and the West: A Realistic Approach*. Translated by Houchang Chehabi. Pp. xii, 228. New York: Norton, 1986. $17.95.

LIEBENOW, J. GUS. *Liberia: The Quest for Democracy*. Pp. xiii, 336. Bloomington: Indiana University Press, 1987. $35.00. Paperbound, $12.95.

LODGE, JULIET. *European Union: The European Community in Search of a Future*. Pp. xviii, 239. New York: St. Martin's Press, 1986. $27.50.

LUNARDINI, CHRISTINE A. *From Equal Sufferage to Equal Rights: Alice Paul and the National Woman's Party, 1910-1928*. Pp. xx, 230. New York: New York University Press, 1986. Distributed by Columbia University Press, New York. $35.00.

MACKENZIE, G. CALVIN, ed. *The In-and-Outers: Presidential Appointees and Transient Government in Washington*. Pp. 239. Baltimore, MD: Johns Hopkins University Press, 1987. $26.50.

MANDLE, JAY R. *Big Revolution, Small Country: The Rise and Fall of the Grenada Revolution*. Pp. xi, 107. Lanham, MD: North-South, 1985. Paperbound, no price.

MARCUS, GEORGE E. and MICHAEL M.J. FISCHER. *Anthropology as Cultural Critique: An Experimental Moment in the Human Sciences*. Pp. xiii, 205. Chicago: University of Chicago Press, 1986. Paperbound, $9.95.

MARIGER, RANDALL P. *Consumption Behavior and the Effects of Government Fiscal Policies*. Pp. 265. Cambridge, MA: Harvard University Press, 1986. No price.

MASON, DAVID S. *Public Opinions and Political Change in Poland, 1980-1982*. Pp. xii, 275. New York: Cambridge University Press, 1985. $37.50.

MITCHELL, BARBARA. *The Practical Revolutionaries: A New Interpretation of French Anarchosyndicalists*. Pp. 314. Westport, CT: Greenwood Press, 1987. $37.95.

MONTICONE, RONALD C. *The Catholic Church in Communist Poland*. Pp. viii, 227. Boulder, CO: East European Monographs, 1986. Distributed by Columbia University Press, New York. $25.00.

MUGHAN, ANTHONY. *Party and Participation in British Elections*. Pp. vi, 159. New York: St. Martin's Press, 1986. $29.95.

MUZUMDAR, HARIDAS T. *Asian Indians' Contributions to America*. Pp. vii, 74. Little Rock, AR: Gandhi Institute of America, 1986. Paperbound, $7.50.

NATHAN, K. S. and M. PATHMANATHAN, eds. *Trilateralism in Asia: Problems and Prospects in U.S.-Japan-Asean Relations*. Pp. xviii, 205. Honolulu: University of Hawaii Press, 1986. $18.00. Paperbound, $12.00.

NEUSNER, JACOB. *Self-Fulfilling Prophecy: Exile and Return in the History of Judaism*. Pp. 230. Boston: Beacon Press, 1987. $25.00.

NICE, DAVID C. *Federalism: The Politics of Intergovernmental Relations*. Pp. xi, 226. New York: St. Martin's Press, 1986. $32.50.

NOLAN, RIALL W. *Bassari Migrations: The Quiet Revolution*. Pp. xv, 199. Boulder, CO: Westview Press, 1986. Paperbound, $22.50.

PADGETT, STEPHEN and TONY BURKETT. *Political Parties and Elections in West Germany: The Search for a New Stability*. Pp. xi, 308. New York: St. Martin's Press, 1987. $32.50. Paperbound, $14.95.

PARSONS, HOWARD L. *Christianity Today in the USSR*. Pp. x, 199. New York: International Publishers, 1987. Paperbound, $6.95.

PEDERSON, WILLIAM D. and ANN M. McLAURIN, eds. *The Rating Game in American Politics: An Interdisciplinary Approach*. Pp. x, 410. New York: Irvington, 1987. Paperbound, $17.95.

PERNICK, MARTIN S. *A Calculus of Suffering: Pain, Professionalism, and Anesthesia in Nineteenth-Century America*. Pp. 421. New York: Columbia University Press, 1987. Paperbound, $14.50.

PERRY, THOMAS L. and JAMES G. FOULKS, eds. *End the Arms Race: Fund Human Needs*. Pp. 336. Seattle: University of Washington Press, 1987. Paperbound, $9.95.

POLLARD, ROBERT A. *Economic Security and the Origins of the Cold War, 1945-1950*. Pp. 378. New York: Columbia University Press, 1987. $14.50.

POSEN, BARRY R. *The Sources of Military Doctrine: France, Britain, and Germany between the World Wars*. Pp. 283. Ithaca, NY: Cornell University Press, 1986. Paperbound, $13.20.

POZO, SUSAN, ed. *Essays on Legal and Illegal Immigration*. Pp. v, 128. Kalamazoo, MI: W. E. Upjohn Institute for Employment Research, 1986. $13.95. Paperbound, $8.95.

REED, ADOLPH, Jr., ed. *Race, Politics, and Culture: Critical Essays on the Radicalism of the 1960s*. Pp. xii, 287. Westport, CT: Greenwood Press, 1986. $35.00.

REYNOLDS, V. et al., eds. *The Sociology of Ethnocentrism: Evolutionary Dimensions of Xenophobia, Discrimination, Racism, and Nationalism*. Pp. xx, 327. Athens: University of Georgia Press, 1987. $40.00.

ROCKAWAY, ROBERT A. *The Jews of Detroit: From the Beginning, 1762-1914*. Pp. xi, 162. Detroit, MI: Wayne State University Press, 1986. $15.95.

ROSS, STEVEN J. *Workers on the Edge: Work, Leisure, and Politics in Industrializing Cincinnati, 1788-1890*. Pp. 406. New York: Columbia University Press, 1987. Paperbound, $17.50.

RUSHDIE, SALMAN. *The Jaguar Smile: Nicaraguan Journey*. Pp. 171. New York: Viking, Elisabeth Shifton Books, 1987. $12.95.

SAFARIAN, A. E. and GILLES Y. BERTIN, eds. *Multinationals, Governments and International Technology Transfer*. Pp. 223. New York: St. Martin's Press, 1987. $37.50.

SAN JUAN, E. *Crisis in the Philippines: The Making of a Revolution*. Pp. xv, 264. South Hadley, MA: Bergin and Garvey, 1986. Paperbound, $14.95.

SANDERSON, STEVEN. *The Transformation of Mexican Agriculture: International Structure and the Politics of Rural Change*. Pp. xxii, 324. Princeton, NJ: Princeton University Press, 1986. $42.00 Paperbound, $10.95.

SARTORI, GIOVANNI. *The Theory of Democracy Revisited*. Part 2, *The Classical Issues*. Pp. 542. Chatham, NJ: Chatham House, 1987. Paperbound, $11.95.

SAUNDERS, HAROLD H. *The Other Walls: The Politics of the Arab-Israeli Peace Process*. Pp. xix, 179. Washington, DC: American Enterprise Institute for Public Policy Research, 1985. Paperbound, no price.

SCHOENHALS, KAI P. and RICHARD A. MELANSON. *Revolution and Intervention in Grenada: The New Jewel Move-*

ment, the United States, and the Caribbean. Pp. x, 211. Boulder, CO: Westview Press, 1985. Paperbound, $22.00.

SEALY, ALBERT H. *Macro Blueprint: For Dialogue to Shape Tomorrow's Economy with Enlightened Public Leadership and Corporate Governance.* Pp. viii, 640. New York: Interbook, 1986. $16.95.

SHAPIRO, IAN. *The Evolution of Rights in Liberal Theory.* Pp. 326. New York: Cambridge University Press, 1986. No price.

SHAPIRO, SUSAN P. *Wayward Capitalists: Target of the Securities and Exchange Commission.* Pp. 227. New Haven, CT: Yale University Press, 1987. Paperbound, $9.95.

SHAVER, PHILLIP and CLYDE HENDRICK, eds. *Sex and Gender.* Pp. 328. Newbury Park, CA: Sage, 1987. Paperbound, no price.

SHUCKBURGH, EVELYN. *Descent to Suez: Foreign Office Diaries 1951-1956.* Pp. x, 380. New York: Norton, 1987. $24.95.

SLANY, WILLIAM Z. et al., eds. *Foreign Relations of the United States, 1952-1954.* Vol. 7, part 1, *Germany and Austria.* Pp. xxix, 1233. Washington, DC: Government Printing Office, 1986. No price.

SMITH, DANIEL BLAKE. *Inside the Great House: Planter Family Life in Eighteenth-Century Chesapeake Society.* Pp. 305. Ithaca, NY: Cornell University Press, 1986. Paperbound, $10.95.

SMITH, DONALD L. *Zechariah Chafee, Jr.: Defender of Liberty and Law.* Pp. 355. Cambridge, MA: Harvard University Press, 1986. $25.00.

SOLOMON, ANTHONY M. *The Dollar, Debt, and the Trade Deficit.* Pp. 55. New York: New York University Press, 1987. $12.50.

STANLEY, JOHN L. *From Georges Sorel: Essays in Socialism and Philosophy.* Pp. 388. New Brunswick, NJ: Transaction Books, 1987. Paperbound, $19.95.

STENELO, LARS-GÖRAN. *The International Critic: The Impact of Swedish Criticism of the U.S. Involvement in Vietnam.* Pp. 194. Boulder, CO: Westview Press, 1986. $29.50.

STIMSON, JOHN and ARDYTH STIMSON, eds. *Sociology: Contemporary Readings.* Pp. 402. Itasca, IL: F. E. Peacock, 1987. Paperbound, no price.

STONE, CARL. *Power in the Caribbean Basin: A Comparative Study of Political Economy.* Vol. 5. Pp. ix, 159. Philadelphia: ISHI, 1986. No price.

SULLIVAN, WILLIAM M. *Reconstructing Public Philosophy.* Pp. xiv, 238. Berkeley: University of California Press, 1986. Paperbound, $8.95.

SZAJKOWSKI, BOGDAN, ed. *Marxist Local Governments in Western Europe and Japan: Politics, Economics and Society.* Pp. xvii, 216. Boulder, CO: Lynne Rienner, 1986. $26.50. Paperbound, $11.95.

THOMPSON, DENNIS L. and DOV RONEN, eds. *Ethnicity, Politics, and Development.* Pp. x, 222. Boulder, CO: Lynne Rienner, 1986. $22.50.

TUCKER, ROBERT W. et al. *SDI and U.S. Foreign Policy.* Pp. 126. Boulder, CO: Westview Press, 1987. Paperbound, $19.50.

TURLEY, WILLIAM S., ed. *Confrontation or Coexistence: The Future of Asean-Vietnam Relations.* Pp. ix, 187. Bangkok: Institute of Security and International Studies, 1985. No price.

VAN DER WEE, HERMAN. *Prosperity and Upheaval: The World Economy 1945-1980.* Pp. 621. Berkeley: University of California Press, 1986. $32.50.

VEEN, HANS JOACHIM. *From Brezhnev to Gorbachev: Domestic Affairs and Soviet Foreign Policy.* Pp. 378. New York: St. Martin's Press, 1987. $45.00.

WALLACE, BRIAN F. *Ownership and Development: A Comparison of Domestic and Foreign Firms in Colombian Manufacturing.* Pp. 176. Athens: Ohio University Press, 1987. Paperbound, $12.00.

WALTER, CAROLYN AMBLER. *The Timing of Motherhood.* Pp. vii, 146. Lexington, MA: D. C. Heath, Lexington Books,

1986. $14.95.

WELTENS, BERT, KEES DE BOT, and THEO VAN ELS, eds. *Language Attrition in Progress*. Pp. 224. Dordrecht, Holland: Foris, 1986. Paperbound, no price.

WILEY, NORBERT, ed. *The Marx-Weber Debate*. Pp. 206. Newbury Park, CA: Sage, 1987. Paperbound, $28.00.

YEANDLE, SUSAN. *Women's Working Lives: Patterns and Strategies*. Pp. vii, 232. New York: Methuen, 1985. Paperbound, $15.95.

ZARTMAN, I. WILLIAM, ed. *The 50% Solution: How to Bargain Successfully with Hijackers, Strikers, Bosses, Oil Magnates, Arabs, Russians, and Other Worthy Opponents in This Modern World*. Pp. 552. New Haven, CT: Yale University Press, 1987. $40.00. Paperbound, $14.95.

ZURCHER, LOUIS A., MILTON L. BOYKIN, and HARDY L. MERRITT, eds. *Citizen-Sailors in a Changing Society: Policy Issues for Manning the United States Naval Reserve*. Pp. xiv, 275. Westport, CT: Greenwood Press, 1986. $35.00.

INDEX

Publisher's Note: The following information is printed in accordance with U.S. postal regulations: Statement of Ownership, Management and Circulation (required by 39 U.S.C. 3685). 1A. Title of Publication: THE ANNALS OF THE AMERICAN ACADEMY OF POLITICAL AND SOCIAL SCIENCE. 1B. Publication No.: 026060. 2. Date of Filing: September 30, 1987. 3. Frequency of Issue: Bi-monthly. 3A. No. of Issues Published Annually: 6. 3B. Annual Subscription Price: paper-inst., $60.00, cloth-inst., $78.00; paper-ind., $28.00, cloth-ind., $42.00. 4. Location of Known Office of Publication: 2111 West Hillcrest Drive, Newbury Park, CA 91320. 5. Location of the Headquarters or General Business Offices of the Publishers: 3937 Chestnut Street, Philadelphia, PA 19104. 6. Names and Complete Addresses of Publisher, Editor, and Managing Editor: Publisher: Sara Miller McCune, 689 Kenwood Ct., Thousand Oaks, California 91360; Editor: Richard D. Lambert, 3937 Chestnut Street, Philadelphia, PA 19104. Managing Editor: None. 7. Owner (if owned by a corporation, its name and address must be stated and also immediately thereunder the names and addresses of stockholders owning or holding 1% or more of total amount of stock. If not owned by a corporation, the names and addresses of the individual owners must be given. If owned by a partnership or other unincorporated firm, its name and address, as well as that of each individual must be given.): The American Academy of Political and Social Science, 3937 Chestnut Street, Philadelphia, PA 19104. 8. Known Bondholders, Mortgagees, and Other Security Holders Owning or Holding 1% or More of Total Amount of Bonds, Mortgages or Other Securities: None. 9. For Completion by Nonprofit Organizations Authorized to Mail at Special Rates (Section 423.12, DMM only): Has not changed during preceding 12 months.

	Av. No. Copies Each Issue During Preceding 12 Months	Actual No. of Copies of Single Issue Published Nearest to Filing Date
10. Extent and Nature of Circulation		
A. Total no. copies printed (net press run)	7557	7127
B. Paid circulation:		
1. Sales through dealers and carriers, street vendors and counter sales	704	361
2. Mail subscription	5217	5166
C. Total paid circulation (sum of 10B1 and 10B2)	5921	5527
D. Free distribution by mail, carrier or other means: samples, complimentary, and other free copies	67	76
E. Total distribution (sum of C and D)	5988	5603
F. Copies not distributed:		
1. Office use, left over, unaccounted, spoiled after printing	1569	1524
2. Return from news agents	0	0
G. Total (sum of E, F1 and 2—should equal net press run shown in A)	7557	7127

11. I certify that the statements made by me above are correct and complete. (Signed) Sara Miller McCune, Publisher.